Numerical Methods for Large-Scale Linear Time-Varying Control Systems and related Differential Matrix Equations

Dissertation

submitted to the

Faculty of Mathematics at **Technische Universität Chemnitz**

in accordance with the requirements for the degree

doctor rerum naturalium (Dr. rer. nat.)

by

Dipl.-Math. techn. Norman Lang

born February 5, 1987 in Schlema, Germany

Advisor: **Prof. Dr. Peter Benner**

Reviewers: **Prof. Dr. Tobias Damm**

Prof. Dr. Tatjana Stykel

submitted: **27th of June, 2017**
defended: **13th of November, 2017**

Bibliografische Information der Deutschen Nationalbibliothek

Die Deutsche Nationalbibliothek verzeichnet diese Publikation in der Deutschen Nationalbibliografie; detaillierte bibliografische Daten sind im Internet über http://dnb.d-nb.de abrufbar.

ISBN 978-3-8325-4700-4

Logos Verlag Berlin GmbH
Comeniushof, Gubener Str. 47,
10243 Berlin
Tel.: +49 (0)30 42 85 10 90
Fax: +49 (0)30 42 85 10 92
INTERNET: http://www.logos-verlag.de

to Nils & Juliane

ACKNOWLEDGEMENTS

Financial Support This thesis was financially supported by the Deutsche Forschungsgemeinschaft DFG in subproject A06 "Model Order Reduction for Thermo-Elastic Assembly Group Models" of the Collaborative Research Center/ Transregio 96 "Thermo-energetic design of machine tools – A systemic approach to solve the conflict between power efficiency, accuracy and productivity demonstrated at the example of machining production".

Personal Thanks First of all, I want to thank my supervisor and teacher Peter Benner. His guidance and constant support were one of the key elements in writing this thesis. Giving the introductory lectures on linear algebra and main parts of my numerical education as undergraduate, Peter Benner opened the doors to the fascinating world of numerical mathematics and finally gave me the opportunity of working in his research group and writing this thesis. Secondly, great thanks goes to my second advisor and good friend Jens Saak who always helped me with a lot of inspiring discussions and consequently a lot of time while struggling with numerous obstacles. Moreover, besides work Jens always had time for good food and not so good movies helping to find a good work-life balance during my time in Magdeburg.

I also like to thank Tobias Breiten, Sara Grundel, Martin Köhler, Patrick Kürschner, Martin Stoll, Matthias Voigt and Heiko Weichelt for being nice colleagues and friends and thus creating a very nice working atmosphere and an even better after-work life.

Special thanks go to André and Judith Schneider with kids for being good friends at work and even more in social life. I enjoyed countless hours being around them.

Nicht zuletzt möchte ich meiner Familie und meinen engsten Freunden für die jahrelange Unterstützung in allen Belangen des täglichen Lebens danken. Besonderer Dank geht hierbei an meine Eltern, die mir in jungen Jahren den richtigen Weg für meine schulische Ausbildung zeigten und mich für die Herausforderungen des Lebens vorbereiteten. Weiter danke ich ganz besonders meiner Freundin Juliane, die jegliche Stimmungsschwankungen bei der Entstehung dieser Arbeit und darüber hinaus in vollem Umfang zu bewältigen wusste. Abschließend danke ich unserem Sohn Nils dafür, dass er ist, wie er ist und damit mein Leben neben der Arbeit bereichert wie niemand sonst.

ABSTRACT

This thesis is concerned with the investigation and development of numerical methods for large-scale linear time-varying (LTV) systems and related differential matrix equations (DMEs). The first major topic concerns a tracking-type finite-time optimal control problem with application to an inverse heat conduction problem. That is, the goal is to compute an input signal that is optimal with respect to the cost for driving the associated system outputs close to some given measurements. It turns out that the main step of this approach is the solution of a differential Riccati equation. Then, the performance of the procedure is shown for a simulation data based test example, as well as for a real-world application relying on real temperature measurements. It is numerically shown that the accuracy of the reconstruction lies within the per mille range.

Secondly, in the context of model order reduction (MOR) for LTV systems, three different approximation approaches are considered, of which two procedures rely on the reduction of linear time-invariant (LTI) approximations of the original LTV model. The first of which uses an approximation of the time-variability by piecewise constant subsystems of LTI structure in terms of switched linear systems that are reduced by the balanced truncation (BT) method. The second scheme is based on a parametric interpretation of the variability and uses parametric MOR (PMOR) in terms of the iterative rational Krylov algorithm (IRKA) applied to local LTI systems in a-priori chosen parameter sample points. Finally, the BT MOR method for LTV systems, mainly based on the solution of the reachability and observability differential Lyapunov equations (DLEs), is investigated with respect to the solution strategy for the DLEs. Moreover, the use of time-varying truncation matrices, as well as a set of time-wise globally truncation bases within the BT approach is investigated. Numerical experiments compare the several MOR techniques with respect to their accuracy and computation times. It is revealed that the BT approach for LTV systems, based on time-varying projection matrices is by far the most expensive approach with respect to the overall computational time but yields the most accurate approximation. Furthermore, its modification, using global truncation bases, can even compete in terms of the online simulation times, while at the same time roughly preserves the accuracy of the original BT approach.

Motivated by the finite-time optimal control problem and the BT method for LTV systems and the associated DMEs, we investigate numerous implicit solvers for ordinary differential equations and their application to the matrix-valued case. Particularly, the rather new class of peer methods is extended to matrix-valued ODEs. For the efficient solution of large-scale DMEs, the well-established classical low-rank Cholesky-type factorization (LRCF) based solution methods are reviewed and further applied to the peer schemes. It is stated that the LRCF based methods, in general, result in complex data and arithmetic for the presented ODE solvers of integration order ≥ 2. Therefore, a low-rank symmetric indefinite factorization (LRSIF), based on a three-term decomposition of the form LDL^T, is proposed for the application to the time integration methods. This decomposition allows to avoid the complex data and arithmetic and is therefore predicted to outperform the classical LRCF schemes. A numerical comparison of the LRCF and LRSIF based methods emphasizes the theoretically predicted superiority of the integration schemes based on symmetric indefinite decomposition.

ZUSAMMENFASSUNG

Die vorliegende Arbeit beschäftigt sich mit der Untersuchung und Entwicklung numerischer Methoden für große, linear zeitvariante Systeme und den damit in Zusammenhang stehenden Differentialmatrixgleichungen. Zunächst wird ein zeitendliches Optimalsteuerungsproblem vom Nachführungstyp, mit Anwendung auf ein inverses Wärmeleitungsproblem, vorgestellt. Hierbei ist es das Ziel, das Eingangssignal, welches optimal im Sinne der Kosten zur Nachführung der zugehörigen Systemausgänge, bezüglich gegebener Messdaten, sein soll, zu berechnen. Die Hauptaufgabe dabei ist das Lösen einer Differential-Riccati-Gleichung. Die Leistungsfähigkeit des Verfahrens ist anhand eines Simulationsdaten basierten Beispiels sowie einer realen Anwendung, basierend auf echten Temperaturmessdaten, dargestellt. Die numerischen Beispiele zeigen, dass die Rekonstruktionsgüte im Promillebereich liegt.

Darauffolgend werden drei verschiedene Verfahren der Modellordnungsreduktion (MOR) für lineare, zeitvariante Systeme betrachtet. Dabei stützen sich zwei der Ansätze auf die Reduktion von linear zeitinvarianten Näherungen des ursprünglichen linear zeitvarianten Modells. Der erste Ansatz verwendet eine Näherung der Zeitveränderlichkeit mittels stückweise konstanter zeitinvarianter Subsysteme im Sinne linearer, geschalteter Systeme, welche mit Hilfe der Methode des balancierten Abschneidens reduziert werden. Der zweite Zugang basiert auf einer parametrischen Interpretation der Zeitveränderlichkeit und stützt sich auf parametrische MOR unter Verwendung des iterativen, rationalen Krylovalgorithmus. Dieser wird in zuvor ausgewählten Parameterstützstellen auf lokal, zeitinvariante Systeme angewendet. Zuletzt wird die Methode des balancierten Abschneidens für zeitvariante Systeme, hinsichtlich der Lösungsstrategie, der im Kern zu lösenden Erreichbarkeits- sowie Beobachtbarkeits-Differential-Lyapunov-Gleichungen, untersucht. Darüber hinaus wird die Anwendung von zeitvarianten sowie zeitlich globalen Reduktionsmatrizen untersucht. Die verschiedenen MOR-Ansätze werden anhand numerischer Beispiele hinsichtlich ihrer Genauigkeit sowie Rechenzeiten verglichen. Dabei zeigt sich, dass der Ansatz des balancierten Abschneidens für linear zeitvariante Systeme, unter Verwendung zeitlich veränderlicher Reduktionsmatrizen, mit Abstand die aufwändigste Methode, bezüglich der Gesamtrechenzeit, darstellt. Andererseits können mit dieser MOR-Strategie die besten Ergebnisse im Bezug auf die Genauigkeit erreicht werden. Außerdem zeigt sich, dass unter Verwendung der zeitlich konstanten Reduktionsmatrizen, Onlinesimulationszeiten im Bereich der anderen MOR-Verfahren erreicht werden können, wobei die Genauigkeit des ursprünglichen BT-Ansatzes nahezu erhalten bleibt.

Motiviert durch die zu lösenden Differentialgleichungen vom Riccati- und Lyapunov-Typ, welche aus dem zeitendlichen Optimalsteuerungsproblem bzw. der Methode des balancierten Abschneidens für LTV Systeme resultieren, werden zahlreiche implizite Löser für gewöhnliche Differentialgleichungen untersucht und auf den matrixwertigen Fall angewendet. Dabei wird insbesondere die recht neue Klasse der Peer-Methoden auf den Fall matrixwertiger Differentialgleichungen erweitert. Im Hinblick auf effiziente Lösbarkeit für große Differentialgleichungen betrachten wir die fest etablierten Lösungsmethoden, basierend auf einer Niedrigrangzerlegung der Form ZZ^T. Es wird herausgestellt, dass die vorgestellten Lösungsverfahren, mit einer Integrationsordnung ≥ 2, auf komplexwertige Daten führen und damit komplexwertige Rechenoperationen erfordern. Aus diesem Grund wird eine symmetrisch, indefinite Zerlegung der Form LDL^T für die verwendeten Zeitintegrationsverfahren vorgestellt. Diese Zerlegung erlaubt es, die komplexen Daten und damit komplexwertige Operationen zu vermeiden. Daraus ergibt sich ein deutlicher Vorteil dieser Zerlegungsstrategie. Ein numerischer Vergleich beider Zerlegungsansätze unterstreicht die bereits theoretisch vorhergesagte Überlegenheit der Integrationsverfahren in der symmetrisch, indefiniten Darstellung.

CONTENTS

LIST OF FIGURES

LIST OF TABLES

LIST OF ALGORITHMS

ACRONYMS AND NOTATION

Acronyms and Abbreviations

ADI	alternating directions implicit
AME	algebraic matrix equation
ARE	algebraic Riccati equation
BDF	Backward differentiation formula
BDF(s)	BDF method of order (s)
BT	balanced truncation
DAE	differential-algebraic equation
DME	differential matrix equation
DIRK	diagonally implicit Runge-Kutta
EKSM	extended Krylov subspace method (aka KPIK)
(F/G)DRE	(filter/generalized) differential Riccati equation
FEM	finite element method
FOM	full-order model
(G)CALE	(generalized) continuous-time algebraic Lyapunov equation
(G)CARE	(generalized) continuous-time algebraic Riccati equation
(G)DLE	(generalized) differential Lyapunov equation
HJE	Hamilton-Jacoby equation
HSV	Hankel singular values
IRKA	iterative rational Krylov algorithm
IVP	initial value problem
KPIK	Krylov-Plus-Inverted-Krylov (aka EKSM)
LRCF	low-rank Cholesky-type factorization
LRSIF	low-rank symmetric indefinite factorization
LTI	linear time-invariant
LTV	linear time-varying
LQ(R/G)	linear-quadratic (regulator/Gaussian)
Mid	midpoint rule
MIMO	multiple-input and multiple-output

MATLAB®	software from The MathWorks Inc.
(P)MOR	(parametric) model order reduction
mRosPeer(s)	modified RosPeer(s)
ODE	ordinary differential equation
PDE	partial differential equation
Peer(s)	implicit peer integration method of order (s)
POD	proper orthogonal decomposition
RKSM	rational Krylov subspace methods
ROM	reduced-order model
Ros(s)	Rosenbrock integration method of order (s)
RosPeer(s)	Rosenbrock-type Peer integration method of order (s)
SISO	single-input and single-output
SVD	singular value decomposition
Trap	trapezoidal rule

Notation

$\mathbb{C}$	field of complex numbers
$\mathbb{C}_+$, $\mathbb{C}_-$	open right/open left complex half plane
$\mathbb{R}$	field of real numbers
$\mathbb{R}_+$, $\mathbb{R}_-$	strictly positive/negative real line
$\mathbb{R}^n$, $\mathbb{C}^n$	vector space of real/complex n-tuples
$\mathbb{R}^{m\times n}$, $\mathbb{C}^{m\times n}$	real/complex $m \times n$ matrices
$\mathbb{Z}$	field of integers
$\lvert\xi\rvert$	absolute value of real or complex scalar ξ
$\jmath$	imaginary unit ($\jmath^2 = -1$)
$\mathrm{Re}(A)$, $\mathrm{Im}(A)$	real and imaginary part of a complex quantity $A = \mathrm{Re}(A) + \jmath\,\mathrm{Im}(A) \in \mathbb{C}^{n\times m}$
$\overline{A}$	$:= \mathrm{Re}(A) - \jmath\,\mathrm{Im}(A)$, complex conjugate of $A \in \mathbb{C}^{n\times m}$
a_{ij}	the i, j-th entry of A
A^T	the transpose of A
A^H	$:= (\overline{a}_{ij})^T$, the conjugate transpose
A^{-1}	inverse of nonsingular A
A^{-T}, A^{-H}	inverse of A^T/A^H
I_n	identity matrix of dimension n
$\Lambda(A)$, $\Lambda(A, E)$	spectrum of matrix A/matrix pair (A, E)
$\lambda_j(A)$, $\lambda_j(A, E)$	j-th eigenvalue of $A/(A, E)$
$\rho(A, E)$	$:= \max\limits_j \lvert\lambda_j(A, E)\rvert$, spectral radius of (A, E)
$\sigma_{\max}(A)$	largest singular value of A
$\mathrm{tr}(A)$	$:= \sum_{i=1}^n a_{ii}$, trace of A

$\|u\|_\infty$	the maximum norm, i.e., the maximum absolute value of the components of a vector u
$\|A\|_F$	$:= \sqrt{\sum_{i,j} a_{ij}^2} = \sqrt{\mathrm{tr}(A^H A)}$, the Frobenius-norm of a matrix $A \in \mathbb{R}^{m \times n}$
$\|u\|$, $\|A\|$	Euclidean vector-, or subordinate matrix norm $\|\cdot\|_2$
$L_2([a,b];\mathbb{R}^n)$	set of square-integrable functions $x(t)$ that satisfy $\int_a^b \|x(t)\|^2\,dt < \infty$
$PC([a,b];\mathbb{R}^n)$	set of piecewise continuous functions $x(t) \in \mathbb{R}^n$, $t \in [b,b]$
$A \succ (\succeq) 0 \,/\, A \prec (\preceq) 0$	short form for A is positive/negative (semi)definite, also abbreviated by *sp(s)d* and *sn(s)d*
$A \succ B$, $A \succeq B$	$:\Leftrightarrow A - B > 0 \,/\, A - B \geq 0$
$A \otimes B$	the Kronecker product of A and B
$\mathrm{vec}(A)$	vectorization operator applied to matrix A
$\dot{f} := \partial_t f := \frac{\partial}{\partial t} f$	first derivative of f with respect to time
$\ddot{f} := \frac{\partial^2}{\partial t^2} f$	second derivative of f with respect to time
$\nabla f := (\partial_1 f, \ldots, \partial_n f)^T$	the gradient of f
$\Delta f := \sum_{i=1}^{n} \partial_i^2 f$	the Laplacian operator applied to f
$\lfloor x \rfloor := \mathrm{floor}(x)$	$:= \max\{k \in \mathbb{Z} \mid x \geq k\}$, the floor function (Gauß brackets) rounding x to the greatest preceding integer $k \in \mathbb{Z}$
$a \overset{!}{\triangle} b$, $\triangle \in \{=, >, <, \geq, \leq\}$	determine a such that the relation $\triangle$ to b holds

CHAPTER 1

INTRODUCTION

Contents

1.1. Motivation

This thesis is devoted to the development and improvement of efficient algorithms for linear control systems and in particular for the subclass of linear time-varying (LTV) systems. Almost every real-world application can be modeled by partial differential equations (PDEs) and/or ordinary differential equations (ODEs). Those being defined with linear dynamics can in a second step be reformulated in terms of linear control systems. If in addition the equations, describing the dynamics are varying over time, one faces the class of LTV control systems. Although the scope of linear control systems for the representation of real-life problems is rather limited, they play an important role in many applications, such as circuit and micro-electro-mechanical systems simulation, mechanical engineering, and many more. Other very interesting and widely spread fields of applications concern nonlinear time-invariant models that, considering a linearization around a time-varying reference trajectory, also result in linear time-varying systems, approximating the original nonlinearity. Detailed elaborations on linear systems can be found in many text books, see e.g. [48, 146, 112, 49]. Due to the importance and complexity of the linear models, the mathematical study and numerical analysis is a highly active research topic. Here, the expression *complexity* is meant to denote the fact that the creation of sufficiently accurate models, representing real-life processes with possibly time-dependent dynamics, easily ends up in a large number of equations. As a direct consequence any computer-aided procedure applied to these so-called *large-scale* systems demands for highly efficient numerical algorithms with respect to the computational effort and the required storage amount.

Two of the most important research areas for linear systems are linear-quadratic (LQ) optimal control and model order reduction (MOR). In general, LQ control is concerned with operating a linear control system at minimal cost, described by the minimization of a certain cost functional, also called *performance index*. In the literature, numerous variations of these minimization objectives, undoubtedly dedicated to specific purposes, are present. A very popular role plays the so-called *tracking problem*, whose particular aim is to make a certain linear combination of the state behave in the same way as a desired target item. Another major distinguishing feature of LQ problems is their definition on either a finite or infinite time line. For details on optimal control and the LQ design, we refer to standard textbooks, such as [146, 147, 143]. However, apart from the usually intended use, the tracking-type LQ design allows the application to linear inverse problems. Considering dynamical processes originating from, e.g., a PDE, the linear system to be solved easily reaches unacceptable complexity in terms of the system dimension, being determined by the spatial and time-wise discretization. Thus, instead of solving a minimization problem arising from a typically applied regularization technique to these systems, an LQ optimal control problem can be formulated. Most of the classical feedback-based optimal control approaches, see, e.g., [184, 143] can be solved using a certain feedback law requiring the solution of a *differential* or *algebraic Riccati equation* (DRE/ARE). Furthermore, even more sophisticated modern $\mathcal{H}_2$- and $\mathcal{H}_\infty$-controller strategies, taking robustness of the control law into account [76, 86, 169], boil down to the solution of a DRE or ARE. Consequently, from the perspective of the computational effort, the main issue is to deal with these large-scale matrix equations. In particular, considering LTV control systems or even LTI systems defined on a finite-time horizon, the associated optimal control problem requires the solution of a nonlinear matrix-valued differential Riccati equation.

Dealing with large-scale dynamical systems, an efficient simulation is only feasible, if the number of unknowns is drastically reduced, while at the same time, the accuracy of the resulting surrogate model can, up to a certain tolerance, be preserved. Here, model order reduction comes into play. MOR for linear systems aims at the approximation of the original large-scale system by a low-dimensional surrogate model. The theory is an already well-established research field [6, 8, 28, 78]. Still there exist open problems, in particular for LTV systems, that are worth studying. To the best of the authors knowledge, in the literature, there exist four basic MOR strategies for LTV systems. The first strategy transforms the LTV system into an LTI or even a sequence of LTI systems [105, 161, 195]. Then classical MOR approaches for LTI systems can be used for the computation of an appropriate reduced-order model

(ROM). For the second ansatz, the time-variability is in a certain sense represented by a parameter-dependency such that parametric MOR (PMOR) techniques [21] can be applied, see, e.g., [74, 75]. The third strategy uses extensions of the MOR counterparts for LTI systems. In this respect the balanced truncation (BT) method has received most attention [183, 202, 144, 128, 171, 68, 50, 51, 201]. Although being theoretically well studied, the BT method for large-scale LTV systems misses an implementable algorithm. This is due to the fact that the BT approach for LTV systems requires the solution of two matrix-valued *differential Lyapunov equations* (DLEs) that for large-scale systems is a computational expensive demand. Some of the above mentioned articles even consider the case of *differential Lyapunov inequalities*. Finally, the fourth approach is the so-called *proper orthogonal decomposition* (POD) method. The POD method first appeared in [123] and is, e.g., also known as *principle component analysis* (PCA) or *Karhunen-Loevĕ expansion* and can be found in many text books as, e.g., [6, 28]. In contrast to the previous procedures, POD is a simulation driven MOR technique based on snapshots taken from solution trajectories generated by a number of forward simulations. Roughly speaking, the central tool is the *singular value decomposition* (SVD) applied to a matrix consisting of the concatenated simulation snapshots. That is, the POD method does not take into account the specific structure of the underlying dynamical system and therefore can similarly be applied to LTI and LTV, or even nonlinear control systems.

As already stated, one of the key tasks in both fields of interest is the solution of matrix-valued equations of differential and/or algebraic nature. In case of the differential matrix equations (DMEs) the first question is to find an appropriate solution strategy. In the literature a lot of work has been put into the solution of DREs, see, e.g., [55, 114, 125, 148, 136, 99]. For most control systems, fast and slow modes are present, and therefore the DMEs related to these systems are often observed to be fairly stiff. Thus integration methods capable of dealing with that stiffness are required. This directly demands for implicit solution strategies, as e.g., presented in [52, 59]. Still, most of these methods rely on the application of dense solution methods to these matrix equations and therefore result in a solution of the size of the problem dimension squared. For large-scale problems, this requires a large amount of computational resources. Therefore modern and efficient algorithms rely on low-rank based solution algorithms. Considering linear control systems with only few inputs and outputs, these are motivated by the experimentally observed and theoretically investigated rapid decay of the singular values and therefore inherent low numerical rank of the solutions. In [149, 30, 31] classical implicit time integration methods, originally developed for standard scalar and vector-valued ODEs, exploiting the low-rank phenomena were presented for the application to large-scale DREs. All these implicit methods lead to the solution of either algebraic Riccati or Lyapunov equations. The most popular solution methods for these algebraic matrix equations are the alternating directions implicit (ADI) iteration and the Krylov subspace methods. For both types sophisticated low-rank formulations are available, see [23, 24, 25, 126] and [185, 60, 61, 194], respectively.

However, allowing the application to large-scale DREs, still these methods suffer from the appearance of complex data for methods of an integration order ≥ 2. In this thesis a low-rank decomposition based on a symmetric indefinite factorization that can overcome the problems arising within the classical low-rank scheme, is discussed. Recently in [98, 190, 192], splitting methods have been applied to large-scale DREs. Exploiting the advantages of the aforementioned symmetric indefinite decomposition, these methods lead to very promising results with respect to the accuracy and computational times of the solvers. Also, methods with high integration orders seem to be easily applicable to the DREs.

1.2. Outline of the Thesis

In Chapter 2 the mathematical foundations, used throughout the thesis, are stated. At first, this includes some basic but important control-theoretic properties of linear dynamical control systems, as well as the definitions of the related, and for this thesis essential, matrix equations. Further, the fundamental concepts of the ADI and Krylov subspace methods for the solution of the algebraic matrix equations (AMEs) are given. As a second part, the basic concept of projection based model order reduction is reviewed. Moreover, the method of balanced truncation for LTI systems and interpolatory model order reduction based on Krylov subspace projection methods are introduced for their use in Chapter 4. Based on the introductory statements to the Hamilton-Jacobi theory also given in Chapter 2, in Chapter 3 a tracking-type LQ regulator (LQR) problem is applied to the solution of an inverse heat conduction problem (IHCP). There, the goal is to reconstruct the thermal input being responsible for a certain system response given in terms of temperature measurements. The procedure is first tested for computed simulation data and secondly applied to a simplified real-world example relying on measurements taken within a real experiment. As a main task, the associated differential Riccati equation has to be solved. Chapter 4 is concerned with the application of model order reduction to linear time-varying systems. Two major strategies are investigated. The first of which is based on an approximation of the LTV system by different LTI system interpretations. A switched linear systems (SLS) representation, as well as parametric formulation of the time variability, are considered. The second major approach presents a practical implementation of the BT method for LTV systems based on the efficient solution of the associated differential Lyapunov equations. Motivated by the differential matrix equations arising in Chapters 3 and 4, a number of standard implicit time integration methods are reviewed in Chapter 5. Additionally, the rather new class of peer methods is introduced for their application to the matrix-valued case. The first part of Chapter 6 presents the classical low-rank Cholesky-type DME solution algorithms, whereas the second part of Chapter 6 introduces the reformulations of these methods based on the symmetric indefinite decomposition. Numerical experiments verifying the results are included to the Chapters 3, 4 and 6 themselves. Finally, Chapter 7 gives a summary of the presented contents together with an overview for future research perspectives.

1.3. Related own publications

Large parts of this thesis have previously been published in the peer reviewed journal articles listed below.

Chapter 3, in particular Section 3.2, includes

> [134]: Norman Lang, Jens Saak, Peter Benner, Steffen Ihlenfeldt, Steffen Nestmann, Klaus Schädlich: *Towards the identification of heat induction in chip removing processes via an optimal control approach*, Production Engineering Research and Development, 9, pp. 343-349, 2015.

Section 4.1 of Chapter 4 is a revised and extended version of

> [133]: Norman Lang, Jens Saak, Peter Benner: *Model Order Reduction for Systems with Moving Loads*, at-Automatisierungstechnik, 62(7), pp. 512-522, 2014.

Section 4.2 of Chapter 4 has mainly been published in

> [135]: Norman Lang, Jens Saak, Tatjana Stykel: *Balanced truncation model reduction for linear time-varying systems*, Mathematical and Computer Modeling of Dynamical Systems, 22(4), pp. 267-281, 2016.

Chapter 5 contains

> [132]: Norman Lang, Hermann Mena, Jens Saak: *On the benefits of the LDL^T factorization for large-scale differential matrix equation solvers*, Linear Algebra and its Applications, 480, pp. 44-71, 2015.

1.4. Used Software and Hardware

The numerical experiments throughout this thesis have been executed on a 64bit CentOS 5.5 system with two Intel® Xeon® X5650@2.67 GHz with a total of 12 cores and 48GB main memory, being one computing node of the linux cluster otto[1] at the *Max Planck Institute for Dynamics of Complex Technical Systems* in Magdeburg. The numerical algorithms have been implemented and tested in MATLAB® version 8.0.0.783 (R2012b). The finite element discretization and the associated model realizations of the examples in Sections 3.3 and 4.3.1 were set up in the finite element software environment FEniCS[2] version 1.6.

[1]http://www.mpi-magdeburg.mpg.de/1012477/otto
[2]https://fenicsproject.org/

CHAPTER 2

MATHEMATICAL BASICS

Contents

2.1. Linear Systems and Optimal Control

This section states a brief review on the essential ideas and most important properties for linear dynamical systems in the context of control theory, as well as the basic concepts for linear-quadratic optimal control problems.

A linear dynamical system of the form

$$\begin{aligned} E\dot{x}(t) &= Ax(t) + Bu(t), \\ y(t) &= Cx(t) + Du(t), \\ x(0) &= x_0, \end{aligned} \tag{2.1}$$

is called a *generalized state-space system*. Here $x \in \mathcal{X} \subset \mathbb{R}^n$ denotes the *state* variable, $u \in \mathcal{U} \subset \mathbb{R}^m$ is the system *input* (or *control*), and $y \in \mathcal{Y} \subset \mathbb{R}^q$ is the *output* of the system, whereas the spaces $\mathcal{X}$, $\mathcal{U}$, $\mathcal{Y}$ are the state, input and output spaces, respectively. The so-called *system matrices* $E, A \in \mathbb{R}^{n \times n}$,$B \in \mathbb{R}^{n \times m}$, $C \in \mathbb{R}^{q \times n}$ and $D \in \mathbb{R}^{q \times m}$ denote the *mass* matrix, the *state-space* matrix, the *input map*, the *output map*, and the *feedthrough map*, respectively. Further, the first equation is the so-called *state equation*, the second is called *output equation*, and the third defines the *initial condition*. As we have $D = 0$ in all the applications considered throughout this thesis, we will neglect the feedthrough matrix for the remainder. Given $m = q = 1$, system (2.1) is called *single-input single-output* (SISO) and *multi-input multi-output* (MIMO) if both are strictly larger than 1. If the matrices E, A, B, C are time independent, the system is called *linear time-invariant* (LTI) and time dependent system matrices $E(t), A(t), B(t), C(t)$ with $E, A : \mathbb{R} \to \mathbb{R}^{n \times n}$, $B : \mathbb{R} \to \mathbb{R}^{n \times m}$, and $C : \mathbb{R} \to \mathbb{R}^{q \times n}$ define a so-called *linear time-varying* system. Systems with $E \equiv I$, we refer to as *standard state-space systems*. Note that for the remainder we assume $E(t), A(t), B(t), C(t)$ to be continuous and bounded matrix-valued functions. Frequently, we write E, A instead of $E(t), A(t)$, etc. if no cause of confusion exists. For a singular matrix E, system (2.1) defines a set of *differential algebraic equations* (DAEs). The case E being non-singular will play a major role throughout the entire thesis. Thus, unless stated otherwise, we assume that the inverse of E exists.

In the remainder, an LTI system is referred to as (E, A, B, C) while for an LTV system we use $(t; E, A, B, C)$ to highlight the time dependence of the system matrices. Provided that E is non-singular, the generalized system (2.1) can always be transformed to a standard state-space system by substituting $A \leftarrow E^{-1}A$, $B \leftarrow E^{-1}B$. Then, we write (A, B, C) and $(t; A, B, C)$ for a standard LTI and LTV system, respectively. Furthermore, in what follows, the control function $u(.)$, defined on the interval $[t_0, t_f]$, is assumed to be contained in a function space of *admissible controls* $\mathcal{U}_{ad}(t_0, t_f)$. If no further specification is given, we write $\mathcal{U}_{ad}$ instead of $\mathcal{U}_{ad}(t_0, t_f)$. In the literature, $\mathcal{U}_{ad}$ is, in general, chosen to be the set of vector-valued piecewise continuous functions $PC([t_0, \infty]; \mathcal{U})$ or the square-integrable functions $L_2([t_0, \infty]; \mathcal{U})$, where the notion of integration is to be understood in the Lebesgue sense. For the sake of notation, we consider $\mathcal{U}_{ad} = PC([t_0, \infty]; \mathbb{R}^m)$ in the remainder. The statements analogously hold for $\mathcal{U}_{ad} = L_2([t_0, \infty]; \mathbb{R}^m)$. In this case the expressions "for all" than, in general, have to be replaced by "a.e.". For an detailed overview on the topic of measure theory, see, e.g., the textbooks [172, 3].

2.1.1. Basic Properties

In control theory, the most important concepts for a linear system (2.1) are based on the *stability*, *reachability*, and *observability* properties.

Before we state the essential control theoretic properties of a linear system (2.1), following [112, Chapter 9], we give basic assumptions on its solvability. Therefore, we consider the homogeneous system

$$\dot{x}(t) = A(t)x(t), \; x(t_0) = x_0. \tag{2.2}$$

Theorem 2.1:
Let $A : \mathbb{R} \to \mathbb{R}^{n \times n}$ be piecewise continuous. Then, the system (2.2) satisfying the Lipschitz condition

$$||A(t)x(t) - A(t)\tilde{x}(t)|| \leq \max ||A(t)|| ||x(t) - \tilde{x}(t)||, \text{ for all } t,$$

has a unique solution $\phi(t; t_0, x_0)$ with initial condition x_0 at $t = t_0$. □

Now, let $\phi_i(., \tau)$ be the solution of (2.2) satisfying $\phi_i(\tau, \tau) = e_i$, where $e_i \in \mathbb{R}^n$ is the ith unit vector. With that, we form the so-called *transition matrix*

$$\Phi_A(., \tau) = [\phi_1(., \tau), \dots, \phi_n(., \tau)]. \tag{2.3}$$

Then, the unique solution to (2.2) is given by

$$x(t) = \Phi_A(t, t_0)x_0. \tag{2.4}$$

Provided that E is non-singular, Equation (2.4) can be extended to inhomogeneous generalized control systems of the form (2.1) with

$$x(u; x_0; t) = \Phi_{E^{-1}A}(t, t_0)x_0 + \int_{t_0}^{t} \Phi_{E^{-1}A}(t, \tau)E^{-1}B(\tau)u(\tau)\, d\tau.$$

Often, for the solution of (2.1) at time t, induced by the input u, we use $x(t)$ instead of $x(u; x_0; t)$.

Given the transition matrix Φ_A, we state the following properties.

Definition 2.2:
Let $A : \mathbb{R} \to \mathbb{R}^{n \times n}$ be continuous and bounded. Then $A(.)$ is an *exponentially stable* evolution if there exist $\rho \geq 1$, and $\delta > 0$ such that

$$||\Phi_A(t, \tau)|| \leq \rho e^{-\delta(t-\tau)}, \text{ for all } t \geq \tau \geq 0.$$ □

For simplicity, we also write $A(.)$ is stable.

Definition 2.3:
Given the system $\Sigma = (t; A, B, C)$,

a) the pair (A,B) is said to be *stabilizable*, if there exists a $K : \mathbb{R} \to \mathbb{R}^{m \times n}$ continuous and bounded, such that $A(.) + B(.)K(.)$ is stable,

b) a state $x_1 \in \mathcal{X}$ is *reachable* from x_0, if there exists an input $u(t)$, of finite energy, and a time $t_1 < \infty$, such that

$$x(u; x_0; t_1) = x_1.$$

c) a state $x(t_0) = x_0 \in \mathcal{X}$ is *controllable* to x_1, if there exists an input $u(t)$ and a time t_1, such that

$$x(u; x_0; t_1) = x_1.$$

d) A state $x \in \mathcal{X}$ is said to be *unobservable* if

$$y = Cx = 0,$$

i.e., if x is indistinguishable from the zero state.

e) A system Σ is *detectable*, if for all solutions x of $\dot{x} = Ax$ with $Cx(t) \equiv 0$, it holds

$$\lim_{t \to \infty} x(t) = 0. \qquad \square$$

Based on Definition 2.3, the following equivalences give some important tools for the analysis of linear systems.

Theorem 2.4:

The system Σ is

a) reachable if and only if the associated *reachability Gramian*

$$P(t_0, t_1) = \int_{t_0}^{t_f} \Phi_A(t_1, \tau) B(\tau) B(\tau)^T \Phi_A(t_1, \tau)^T \, d\tau \tag{2.5}$$

is positive definite with respect to $t_0 < t_1$.

b) observable if and only if the associated *observability Gramian*

$$Q(t_f, t_0) = \int_{t_0}^{t_1} \Phi_A(\tau, t_0)^T C(\tau)^T C(\tau) \Phi_A(\tau, t_0) \, d\tau \tag{2.6}$$

is positive definite with respect to $t_0 < t_1$. $\square$

Remark 2.5:

For continuous-time linear systems, the concepts of reachability and controllability are equivalent. Moreover, the reachability of a system (2.1) implies the stabilizabilty and observability implies the detectability. $\square$

LTI Systems All of the statements below can be found in standard textbooks as, e.g., [11, 49, 146, 188, 101].

For linear time-invariant systems the transition matrix is given by $\Phi_{E^{-1}A}(t,\tau) = e^{E^{-1}A(t-\tau)}$. Thus, it is easy to show that a system (2.1) is stable, if $\Lambda(A,E) \subset \mathbb{C}_-$. Given a stable generalized state-space system (E, A, B, C), the Gramians P, Q from (2.5) and (2.6), respectively, are defined for $t_f = \infty$ and are given by the so-called *infinite reachability* and *observability* Gramians

$$P = \int_0^\infty e^{E^{-1}A\tau}E^{-1}BB^TE^{-T}e^{(E^{-1}A)^T\tau}\,d\tau, \qquad Q = \int_0^\infty e^{(E^{-1}A)^T\tau}C^TCe^{E^{-1}A\tau}\,d\tau,$$

respectively, see, e.g., [6, Section 4.3]. Another, important tool in the analysis of linear time-invariant dynamical systems is the *transfer function*, that can be obtained by applying the *Laplace transformation* to a system (2.1). Then, the transfer function for a zero initial state reads

$$\mathcal{H}(s) = C(sE - A)^{-1}B \in \mathbb{R}^{q\times m}, \quad s \in \mathbb{C}. \tag{2.7}$$

Note that the assumption of a zero initial state is no restriction in the case of linear systems since that can always be obtained by a simple shift of the origin of ordinates. Further, the transfer function directly relates the inputs and outputs in the frequency domain in the form $Y(s) = \mathcal{H}(s)U(s)$ with $Y(s)$ and $U(s)$ being the Laplace transforms of the output $y(t)$ and the input $u(t)$, respectively. Moreover, in the case of SISO systems with B, $C^T \in \mathbb{R}^n$, the transfer function is a scalar rational function with respect to the frequency variable s.

Definition 2.6:

The system

$$\begin{aligned} E^T\dot{w}(t) &= -A^Tw(t) - C^T\tilde{y}(t), \\ \tilde{u}(t) &= B^Tw(t) + D\tilde{y}(t), \\ w(t_f) &= w_{t_f}, \end{aligned} \tag{2.8}$$

is called the *dual* (or *adjoint*) system of (2.1) and is stated in reverse time. □

2.1.2. Related Matrix Equations

Many applications concerning linear dynamical systems of the form (2.1), such as balancing based model order reduction and optimal control problems need to deal with matrix equations. In particular, in this thesis we consider linear time-varying systems (2.1) and therefore, matrix-valued ordinary differential equations are of central interest. Differential matrix equations mainly arise in applications related to time-varying systems $(t; A, B, C)$ and $(t; E, A, B, C)$, but also in, e.g., finite-time optimal control problems for LTI systems (A, B, C) and (E, A, B, C), see, e.g., Section 3.

Differential Riccati Equation

The *differential Riccati equation* is one of the most popular nonlinear matrix-valued ordinary differential equations arising in optimal control, optimal filtering, H_∞ control of LTV systems and many more, see e.g., [2, 106, 108, 160]. In terms of an optimal control problem for standard state-space systems $(t; A, B, C)$, the DRE reads

$$-\dot{X}(t) = A(t)^T X(t) + X(t)A(t) - X(t)S(t)X(t) + W(t) \tag{2.9}$$

with the matrices $S(t) = B(t)B(t)^T \in \mathbb{R}^{n\times n}$, $W(t) = C(t)^T C(t) \in \mathbb{R}^{n\times n}$. Together with the final condition $X(t_f) = X_{t_f}$, Equation (2.9) states an initial value problem (IVP) in reverse time. This reverted IVP can easily be reformulated into a standard IVP by substituting t with $t_f - t + t_0$ and defining the matrices $\tilde{A}(t) = A(t_f - t + t_0)$, $\tilde{S}(t) = S(t_f - t + t_0)$, $\tilde{W}(t) = W(t_f - t + t_0)$ and $\tilde{X}(t) = X(t_f - t + t_0)$. Taking into account that the derivative $\frac{\partial}{\partial t}(X(t_f - t + t_0)) = -\dot{X}(t_f - t + t_0) = -\dot{\tilde{X}}(t)$, we obtain the auxiliary DRE

$$\dot{\tilde{X}}(t) = \tilde{A}(t)^T \tilde{X}(t) + \tilde{X}(t)\tilde{A}(t) - \tilde{X}(t)\tilde{S}(t)\tilde{X}(t) + \tilde{W}(t), \tag{2.10}$$

where the initial condition is denoted by $\tilde{X}_0 = \tilde{X}(t_0) = X(t_f - t_0 + t_0) = X_{t_f}$. Since, the DRE (2.9) can easily be transformed into the DRE (2.10), and vice versa, in the remainder, we will use the notation A, B, C, X instead of $\tilde{A}$, $\tilde{B}$, $\tilde{C}$, $\tilde{X}$. Analogously, for a generalized system (E, A, B, C), from (2.10) with $A(t) \leftarrow E(t)^{-1}A(t)$ and $B(t) \leftarrow E(t)^{-1}B(t)$, we obtain

$$\dot{X}(t) = A(t)^T E(t)^{-T} X(t) + X(t)E(t)^{-1}A(t) - X(t)E(t)^{-1}S(t)E(t)^{-T}X(t) + W(t).$$

Now, substituting $X(t)$ with $E(t)^T X(t)E(t)$, in order to avoid the matrix inversion of $E(t)$, for its time derivative, we have

$$\frac{\partial}{\partial t}\left(E(t)^T X(t)E(t)\right) = \dot{E}(t)^T X(t)E(t) + E(t)^T \dot{X}(t)E(t) + E(t)^T X(t)\dot{E}(t),$$

provided $E(t)$ is continuous differentiable. Then, the generalized DRE (GDRE) is defined as

$$\begin{aligned} E(t)^T \dot{X}(t)E(t) = &\left(A(t) + \dot{E}(t)\right)^T X(t)E(t) + E(t)^T X(t)\left(A(t) + \dot{E}(t)\right) \\ &- E(t)^T X(t)S(t)X(t)E(t) + W(t) \end{aligned} \tag{2.11}$$

and the associated initial condition transforms to $E(t_0)^T X(t_0)E(t_0) = X_0$. That is, the solution of the GDRE (2.11) requires the knowledge of the time derivative of $E(t)$ or an appropriate approximation. Furthermore, for some of the integration methods, presented in Chapter 5, a time-varying matrix $E(t)$ would considerably complicate the procedures. Therefore, if not stated otherwise, in the remainder, we assume E to be constant. That is, $\dot{E} = 0$ and the resulting GDRE and initial condition read

$$\begin{aligned} &E^T \dot{X}(t)E = A(t)^T X(t)E + E^T X(t)A(t) - E^T X(t)S(t)X(t)E + W(t), \\ &E^T X(t_0)E = X_0. \end{aligned} \tag{2.12}$$

However, the DRE (2.10) is uniquely solvable if the following conditions are satisfied.

Theorem 2.7 (compare [2, Theorem 4.1.6]):
Let $S(t)$, $W(t) \geq 0$ be continuous and bounded for $t \geq t_0$, then the unique solution $\tilde{X}$ of (2.10) with initial condition $\tilde{X}_0 = \tilde{X}(t_0) \geq 0$ exists for $t \geq t_0$ with $\tilde{X} \geq 0$. □

With E being non-singular the same result holds for (2.12). For an in-depth study of the theory of the differential Riccati equations see, e.g., [2, 118, 164].

Differential Lyapunov Equation

The *differential Lyapunov equation* is a linear matrix-valued differential equation and can be understood as a special case of the DRE. The differential Lyapunov equations related to the standard system $(t; A, B, C)$ and its generalized version $(t; E, A, B, C)$ are given by

$$\dot{X}(t) = A(t)X(t) + X(t)A(t)^T + W(t), \tag{2.13}$$
$$E\dot{X}(t)E^T = A(t)X(t)E^T + EX(t)A(t)^T + W(t) \tag{2.14}$$

with initial conditions $X(t_0) = X_0$ and $EX(t_0)E^T = X_0$, respectively. The most important DLEs in control theory are the reachability and observabilty DLEs arising in, e.g., balanced truncation model order reduction of linear time-varying systems, see Section 4.2.

Theorem 2.8 (compare [2, Corollary 1.1.6]):
Let $A(.)$, $W(.)$ be continuous and bounded. Then, the unique solution of the differential Lyapunov equation (2.13) exists and is given by

$$X(t) = \Phi_A(t, t_0)X_0\Phi_A(t, t_0)^T + \int_{t_0}^{t} \Phi_A(t, \tau)W(\tau)\Phi_A(t, \tau)^T \, d\tau.$$

□

Again, for non-singular E, Theorem 2.8 also holds for (2.14). From Theorem 2.8, we infer that the solution $X(t) \geq 0$ for $t \geq t_0$ if the initial condition $X_0 \geq 0$ and $W(t) \geq 0$ for $t \geq t_0$.

Besides the differential matrix equations, we also have to deal with their algebraic counterparts. Algebraic matrix equations play a major role in infinite-time optimal control [143], model order reduction applications to LTI systems [6, 28] and within the solution of differential matrix equations [52, 59, 149].

Algebraic Riccati Equation

An *algebraic Riccati equation* is defined by

$$A^T X + XA - XSX + W = 0. \tag{2.15}$$

For E being non-singular, the generalized ARE (GARE) reads

$$A^T XE + E^T XA - E^T XSXE + W = 0. \tag{2.16}$$

Note that in contrast to the DRE with given initial condition, the ARE in general is not uniquely solvable. There may exist infinitely many solutions. Among those, in control theory usually the stabilizing solution X_* is of particular interest. The following theorem specifies several conditions under which a stabilizing solution to the AREs (2.15) and (2.16) related to the control system (2.1) can uniquely be derived.

Theorem 2.9:

If S, $W \geq 0$, and the system (E, A, S, W) is stabilizable and detectable, then there exists a unique symmetric stabilizing solution X_* of the GARE (2.16) such that $\Lambda(A - SX_*E, E) \subset \mathbb{C}_-$. If in addition the system (E, A, S, W) is observable, then $X_* > 0$. □

For a proof and detailed derivation, see, e.g., [130]. It is easy to note that for $E = I$ we get the corresponding result for the ARE (2.15).

Algebraic Lyapunov Equation

A linear, symmetric, algebraic matrix equation

$$AX + XA^T + W = 0 \tag{2.17}$$

is called the *algebraic Lyapunov equation* (ALE) and its generalized version (GALE) reads

$$AXE^T + EXA^T + W = 0. \tag{2.18}$$

Under certain conditions, the ALEs (2.17) and (2.18) admit a unique solution. We again state the result for the generalized case which simplifies to the standard case for $E = I$.

Theorem 2.10 (compare [2, Corollary 1.1.4]):

The generalized algebraic Lyapunov equation (2.18) has a unique solution X if and only if $\lambda_j \neq -\lambda_k$, $\forall \lambda_j, \lambda_k \in \Lambda(A, E)$. This in particular means, that the pencil (A, E) must not have pure imaginary eigenvalues. □

Remark 2.11:

The algebraic matrix equations (2.15), (2.16) and (2.17), (2.18) are often referred to as continuous-time ARE (CARE), GARE (GCARE) and continuous-time ALE (CALE), GALE (GCALE) in order to distinguish them from their discrete versions. Since in this thesis we only consider continuous-time linear dynamical systems, we neglect the additional indicator. □

Note that for $W = W^T$, $S = S^T$ and an initial condition $X_0 = X_0^T$ (final condition $X_{t_f} = X_{t_f}^T$) for the differential equations, the above matrix equations are symmetric and therefore admit a symmetric solution $X = X^T$.

2.1.3. Solution of the Algebraic Riccati Equation

There is a large variety of ARE solvers in the literature, see, e.g., [35, 186] for a more detailed overview. Still, here we restrict to Newton's method for the solution of the nonlinear AREs, see, e.g., [117, 130]. This will serve as the basis for efficient solution strategies for sparse large-scale problems.

The application of Newton's method to the nonlinear matrix-valued ARE

$$\mathcal{R}(X) = A^T X + XA - XSX + W = 0$$

yields the update scheme

$$\begin{aligned} X_{\ell+1} &= X_\ell + N_\ell, \\ \mathcal{R}'|_{X_\ell}(N_\ell) &= -\mathcal{R}(X_\ell), \end{aligned} \tag{2.19}$$

where $\mathcal{R}'|_{X_\ell}$ is the Fréchet derivative of the Riccati operator $\mathcal{R}$ at X_ℓ, represented by the Lyapunov operator

$$\mathcal{R}'|_{X_\ell} : U \to (A - SX_\ell)^T U + U(A - SX_\ell). \tag{2.20}$$

Hence, Newton's method for the ARE results in the solution of an algebraic Lyapunov equation of the form

$$\begin{aligned} X_{\ell+1} &= X_\ell + N_\ell, \\ A_\ell^T N_\ell + N_\ell A_\ell &= -\mathcal{R}(X_\ell) \end{aligned}$$

with $A_\ell = A - SX_\ell$ at every Newton step. In addition, using the reformulations by Kleinmann [117] the procedure can be simplified to

$$A_\ell^T X_{\ell+1} + X_{\ell+1} A_\ell = -W - X_\ell S X_\ell,$$

directly computing the solution $X_{\ell+1}$. The initial value for Newton's method is often chosen to be $X_0 = 0$.

Summarizing, we find that the solution of the AREs by Newton's method is also based on the solution of a number of algebraic Lyapunov equations.

2.1.4. Solution of the Algebraic Lyapunov Equation

Anticipating Chapters 3- 5, the key element, for the solution of differential Riccati and Lyapunov equations and the model order reduction approaches to be presented, will be the solution of algebraic Lyapunov equations. Therefore, consider the ALE

$$AX + XA^T = -W, \quad W = W^T. \tag{2.21}$$

For small-scale problems a number of solution strategies for the more general Sylvester matrix equation have already been stated beginning in the early 1970's. Starting with the Bartels-Stewart algorithm in [13] many modifications occurred in the following years, see e.g., [80, 110, 111]. In [157] a generalization of the Bartels-Stewart algorithm [13] and an extension to Hammarling's method [97], particularly adjusted to the solution of generalized Lyapunov equations, are presented. In [32, 33] the sign function method, originally developed for the solving AREs, was adapted for the solution of GALEs. Methods, further exploiting the specific structure of the Lyapunov equations and current computer architectures, can be found in recent studies [120, 121, 122].

For large-scale problems, in general, the solutions X will be dense even if the system matrices are sparse. Therefore, it is recommended to never explicitly form those matrices. This, in fact, transfers to the time integration methods based on ALE solutions. In most applications the right hand side of Equation (2.21) is given as a product of matrices containing factors of low numerical rank. Thus, the key ingredient towards an efficient solution strategy for large-scale ALEs is the observation that the solution X often is of low numerical rank, too. Many contributions address the question under which conditions a significant singular value decay of X can be observed and therefore a good low-rank approximation can be obtained, see [9, 85, 159, 170, 198, 126]. In this thesis we will distinguish the two-term low-rank based representation $X \approx ZZ^H$ with $Z \in \mathbb{K}^{n \times n_Z}$, $\mathbb{K} \in \{\mathbb{R}, \mathbb{C}\}$, $n_Z \ll n$, which we refer to as the classical *low-rank Cholesky-type factorization (LRCF)* of the solution X and a three-term decomposition of the form $X \approx LDL^T$ with $L \in \mathbb{R}^{n \times n_\ell}$ and $D \in \mathbb{R}^{n_\ell \times n_\ell}$, $n_\ell \ll n$. The latter, we will call a *low-rank symmetric indefinite factorization (LRSIF)*. The first algorithms exploiting the low-rank structure are based on the alternating directions implicit (ADI) iteration using the LRCF ZZ^H [158]. It is clear that for $Z \in \mathbb{R}^{n \times n_Z}$ the solution X to the ALE (2.21) is recovered by $X \approx ZZ^T$. Over the years the ADI iteration for large-scale ALEs has been extensively studied [18, 26, 139, 140, 168]. The most recent improvements can be found in [23, 24, 25]. Meanwhile, additional methods based on standard, [167, 109], extended, [185, 194], and rational, [60, 61, 62], Krylov subspace projections were developed for solving algebraic Lyapunov equations. On the other hand, only a few contributions consider the LRSIF formulation. In the context of solving large-scale matrix equations this approach originally was developed for the application of the ADI iteration to the more general setting of Sylvester equations [27]. Nevertheless, in [131, 132] it has been shown that this factorization is of major importance for the solution of the differential matrix equations (2.12) and (2.14) in order to avoid complex data and arithmetic arising within time integration methods of order $s \geq 2$.

In the following, we give a short sketch of the ADI iteration and Krylov subspace based Lyapunov solvers, in both the classical low-rank factorization and the symmetric indefinite decomposition. A more detailed elaboration on the ALEs arising in applications and the associated advantages and disadvantages of both factorization strategies is given in Chapter 6.

ADI Iteration based Lyapunov Solver

Here, we only give a basic insight to the origin of the ADI and its low-rank version. For the most recent results and a number of very important improvements, we refer to the very sophisticated elaborations in [126].

The ADI iteration was originally developed in [156] for the solution of linear systems

$$Au = b$$

with $A \in \mathbb{R}^{n \times n}$ being symmetric positive definite, representing a finite difference formulation of elliptic and parabolic partial differential equations. If further, A can be expressed as $A = H + V$, where $H, V \in \mathbb{R}^{n \times n}$ are also symmetric positive definite, then there exists a set of parameters $\alpha_j \in \mathbb{R}_+$ and for $j = 1, 2, \ldots$ the ADI iteration can be defined as the two-step iteration

$$\begin{aligned} (H + \alpha_j I)u_{j-\frac{1}{2}} &= (\alpha_j I - V)u_{j-1} + b, \\ (V + \alpha_j I)u_j &= (\alpha_j I - H)u_{j-\frac{1}{2}} + b, \end{aligned} \tag{2.22}$$

and $u_j \to u$ for $j \to \infty$. Provided that the matrices H and V commute, there exists a set of shift parameters α_j such that the ADI iteration (2.22) reaches a superlinear convergence rate, see [17]. However, for A being stable, the Lyapunov equation (2.21) represents an ADI model problem [203]. That is, from noting that the linear operator $L : X \mapsto AX + XA^T$ is the sum of the commuting operators $L_L : X \mapsto AX$ and $L_R : X \mapsto XA^T$ the application of the ADI procedure (2.22) yields the scheme

$$\begin{aligned} (A + \alpha_j I)X_{j-\frac{1}{2}} &= -W - X_{j-1}(A^T - \alpha_j I), \\ (A + \alpha_j I)X_j^H &= -W - X_{j-\frac{1}{2}}^H(A^T - \alpha_j I). \end{aligned} \tag{2.23}$$

that is started with an initial guess $X_0 = X_0^T$. Here, the shift parameters are no longer restricted to be real, i.e., $\alpha_j \in \mathbb{C}_-$. Rewriting the scheme (2.23) into a one-step iteration, we obtain

$$\begin{aligned} X_j = &-2\,\mathrm{Re}\big(\alpha_j\big)\,(A + \alpha_j I)^{-1} W (A + \alpha_j I)^{-H} \\ &+ (A + \alpha_j I)^{-1}(A - \overline{\alpha_j} I)X_{j-1}(A - \overline{\alpha_j} I)^H (A + \alpha_j I)^{-H}. \end{aligned} \tag{2.24}$$

This allows to easily formulate the low-rank versions of the ADI, necessary for the efficient computation of the solution to large-scale ALEs of the form (2.21) and consequently the AREs represented by (2.15). Note that the solution X of (2.21) stays real, if the set of shift parameters α_j is closed under complex conjugation.

Classical low-rank Cholesky-type factorization For the classical low-rank ADI representation, the right hand side W of (2.21) is originally assumed to be given in the form $W = GG^T$ with $G \in \mathbb{R}^{n \times n_G}$, $n_G \ll n$. Hence, the one-step formulation (2.24) becomes

$$X_j = -2\,\mathrm{Re}\big(\alpha_j\big)\,(A+\alpha_j I)^{-1}GG^T(A+\alpha_j I)^{-H} + (A+\alpha_j I)^{-1}(A-\overline{\alpha_j}I)X_{j-1}(A-\overline{\alpha_j}I)^H(A+\alpha_j I)^{-H}.$$

Introducing a low-rank representation $X_j = Z_j Z_j^H$ and $Z_0 = 0$, Equation (2.24) can be stated in terms of the factors Z_j, given as

$$\begin{aligned} Z_1 &= \sqrt{-2\,\mathrm{Re}(\alpha_1)}(A+\alpha_1 I)^{-1}G, \\ Z_j &= \left[\sqrt{-2\,\mathrm{Re}\big(\alpha_j\big)}(A+\alpha_j I)^{-1}G, \quad (A+\alpha_j I)^{-1}(A-\overline{\alpha_j}I)Z_{j-1}\right]. \end{aligned} \tag{2.25}$$

From (2.25) one observes that the solution factors are formed by successively adding a column block of fixed size n_G throughout the iteration. Still, note that in the above representation all columns have to be processed at each iteration step and therefore the process gets increasingly expensive. Following the achievements in [139, 140], the procedure can drastically be simplified. That is, defining $\gamma_j = \sqrt{-2\,\mathrm{Re}\big(\alpha_j\big)}$, $T_j = A - \overline{\alpha_j}I$ and $S_j = (A+\alpha_j I)^{-1}$ and by the observation that these expressions commute, (2.25) can be rewritten in the form

$$Z_i = \big[\gamma_j S_j W, \gamma_{j-1}(S_{j-1}T_j)S_j W, \ldots, \gamma_1(S_1T_2)S_2\cdots(S_{j-1}T_j)S_j W\big]. \tag{2.26}$$

Finally, note that the order of the shift parameters does not affect the computations. Hence, reversing the sequence of the iterates, we obtain the low-rank ADI iteration

$$\begin{aligned} V_1 &= (A+\alpha_1 I)^{-1}G, \qquad Z_1 = \sqrt{-2\,\mathrm{Re}(\alpha_1)}V_1, \\ V_j &= V_{j-1} - (\alpha_j + \overline{\alpha_{j-1}})(A+\alpha_j I)^{-1}V_{j-1}), \\ Z_j &= [Z_{j-1}, \sqrt{-2\,\mathrm{Re}\big(\alpha_j\big)}V_j]. \end{aligned} \tag{2.27}$$

Given this form, only n_G columns need to be processed within every step of the ADI. Still, there remain several questions.

The first crucial question addresses the efficient computation of a stopping criterion. Exploiting the low-rank structure of the iterates within the ADI, the Lyapunov residual can also be stated in terms of a low-rank factorization, see [23, 126]. In particular it can be shown, that the rank of the residual is limited by the rank of the right hand side factor G of (2.21). The result is summarized in the following theorem that provides an efficient strategy to compute the Lyapunov residual that serves as a stopping criterion for the ADI algorithm.

Theorem 2.12 ([24, Theorem 4.1], [126, Theorem 3.5]):

The residual at iteration step j of the classical low-rank ADI iteration is of rank at most n_G and defined by

$$AZ_jZ_j^H + Z_jZ_j^HA^T + GG^T = R_jR_j^H,$$

where $R_j \in \mathbb{C}^{n\times n_G}$ is given by

$$R_j = (A - \overline{\alpha_j}I)V_j = R_{j-1} - 2\,\mathrm{Re}\big(\alpha_j\big)\,V_j \in \mathbb{C}^{n\times n_G} \tag{2.28}$$

with $R_0 = G$. If $\alpha_j \notin \Lambda(A)$ for all j, then the rank is exactly n_G. □

Secondly, note that the shifts are allowed to be complex and therefore the iterates Z_j might also become complex even for real system matrices entering the ALE (2.21). To the best of the author's knowledge, the literature provides three approaches, reducing the amount of complex arithmetic throughout the computations. All of them are based on the assumption that complex shifts always appear in conjugate pairs $(\alpha_j, \alpha_{j+1} = \overline{\alpha_j})$, such that the set of shifts $\mathcal{A} = (\alpha_1, \ldots, \alpha_J)$ is closed under complex conjugation, i.e., $\mathcal{A} = \overline{\mathcal{A}}$. In fact this appears to be a rather natural convention since complex eigenvalues of real matrices also appear in complex conjugate pairs. In [158, 140, 26], the low-rank ADI is formulated in a completely real manner, exploiting the identity

$$(A \pm \alpha I)(A \pm \overline{\alpha} I) = A^2 \pm 2\,\mathrm{Re}(\alpha)\,A + |\alpha|^2 I, \quad \forall A \in \mathbb{R}^{n\times n},\ \alpha \in \mathbb{C}.$$

Although this approach avoids complex arithmetic at all, for large-scale problems the expression A^2 either contaminates the original sparsity of A, such that sparse direct solvers cannot be applied efficiently, or leads to an increasing condition number that might worsen the efficiency of iterative solution strategies. Another approach is provided by the latest low-rank ADI implementation of LyaPack [159] that unfortunately is not documented in the literature. Here, the algorithm uses complex arithmetic, but forms real blocks Z_j by using the real and imaginary parts of the iterates V_j, V_{j+1} associated to the pair $(\alpha_j, \alpha_{j+1} = \overline{\alpha_j})$. In a similar manner, but staying in real arithmetic, the third procedure [23, 126] exploits the fact that the iterate V_{j+1} is explicitly known once V_j has been computed. The result is summarized in the following theorem.

Theorem 2.13 ([126, Theorem 4.2]):

Assume a set of shifts $\mathcal{A} = (\alpha_1, \ldots, \alpha_J)$ with $\mathcal{A} = \overline{\mathcal{A}}$ and real matrices A, G defining the ALE (2.21). Then for two subsequent blocks V_j, V_{j+1} of the low-rank ADI iteration associated with a pair of complex conjugate shifts $(\alpha_j, \alpha_{j+1} = \overline{\alpha_j})$, using the low-rank residual (2.28), it holds

$$\begin{aligned} V_{j+1} &= \overline{V_j} + 2\delta_j\,\mathrm{Im}\big(V_j\big), \\ R_{j+1} &= R_{j-1} - 4\,\mathrm{Re}\big(\alpha_j\big)\big(\mathrm{Re}\big(V_j\big) + \delta_j\,\mathrm{Im}\big(V_j\big)\big) \end{aligned}$$

with $\delta_j = \frac{\mathrm{Re}(\alpha_j)}{\mathrm{Im}(\alpha_j)}$. Furthermore, regardless of the type of shifts encountered before, if α_j is real, then V_j and R_j are real $n \times n_G$ matrices. □

Algorithm 2.1 Classical low-rank factored ADI method

INPUT: $E,\ A,\ W = GG^T, G \in \mathbb{R}^{n\times n_G}$, defining the GALE (2.18), tolerance $0 < \varepsilon \ll 1$.
OUTPUT: $Z \in \mathbb{R}^{n\times n_G}$, such that $ZZ^T \approx X$.

1: $R_0 = V_0 = G,\ \alpha = [],\ \ j = 1$
2: **while** $\|R_{j-1}^T R_{j-1}\|_2 \geq \varepsilon \|G^T G\|_2$ **do**
3: **if** $j > \dim(\alpha)$ **then**
4: Compute shift parameters α based on $\Lambda(\hat{A}, \hat{E})$ with $\hat{E} = \hat{V}_j^T E \hat{V}_j,\ \hat{A} = \hat{V}_j^T A \hat{V}_j$ and $\hat{V}_j = \mathrm{orth}\left(V_{j-1}\right)$
5: **end if**
6: Solve $(A + \alpha_j E)V_j = R_{j-1}$ for V_j.
7: **if** α_j is real **then**
8: $R_j = R_{j-1} - 2\alpha_j E V_j,\ Z_j = [Z_{j-1},\ \sqrt{-2\alpha_j} V_j]$
9: $j = j + 1$
10: **else**
11: $\eta_j = 2\sqrt{-\mathrm{Re}\left(\alpha_j\right)},\ \delta_j = \frac{\mathrm{Re}\left(\alpha_j\right)}{\mathrm{Im}\left(\alpha_j\right)}$
12: $R_{j+1} = R_{j-1} + \eta_j^2 E\left(\mathrm{Re}\left(V_j\right) + \delta_j \,\mathrm{Im}\left(V_j\right)\right)$
13: $Z_{j+1} = [Z_{j-1},\ \eta_j\left(\mathrm{Re}\left(V_j\right) + \delta_j \,\mathrm{Im}\left(V_j\right)\right),\ \eta_j \sqrt{(\delta_j^2 + 1)}\ \mathrm{Im}\left(V_j\right)]$
14: **end if**
15: $j = j + 1$
16: **end while**

Finally, the clever selection of shift parameters, clearly influencing the convergence speed of the ADI iteration, see, e.g., [170, 203], is of interest. In the literature numerous strategies, providing sets of precomputed shifts are presented. For completeness a few approaches, such as the computation of optimal Wachspress shifts [203], the heuristic Penzl shifts [158], or Leja points [189] are mentioned. Although these methods have proven their effectiveness for numerous problems, optimal or high quality shifts, in particular for large-scale problems, are in general difficult to compute. Either, these methods are computationally expensive or require special a-priori knowledge. A rather new idea deals with the shift computation during the runtime of the low-rank ADI iteration. For that, two different approaches are investigated up to now [25, 126]. The first of them uses a projection of the coefficient matrices $A,\ E$ onto subspaces spanned by certain low-rank iterates that are computed within the iteration anyway. The shift parameters are then computed by solving the associated small-scale eigenvalue problem. Several subspaces based on the current iterates V_j, the residual factor R_j or the entire solution factor Z_j are considered. The second strategy is based on a shift computation, aiming at the minimization of the Lyapunov residual norm in each iteration. A thorough derivation of these so-called *self-generating shifts* can be found in [126]. In this thesis, the first ansatz, using the projection subspace based on V_j, is implemented. A sketch of the procedure for GALEs is given in Algorithm 2.1.

Algorithm 2.2 Symmetric indefinite low-rank factored ADI method

INPUT : Matrices A, $W = GSG^T$ defining (2.21), proper set of shift parameters $\{\alpha_1, \ldots, \alpha_{j_{ADI}}\} \subset \mathbb{C}_-$, tolerance $0 < \varepsilon \ll 1$.
OUTPUT: L, D such that $LDL^T \approx X$.

1: $R_0 = V_0 = G,\ j = 1$
2: **while** $\|R_{j-1}^T R_{j-1} S\|_2 \geq \varepsilon \|G^T G S\|_2$ **do**
3:   **if** $j > \dim(\alpha)$ **then**
4:     Compute shift parameters α based on $\Lambda(\hat{A}, \hat{E})$ with $\hat{E} = \hat{V}_j^T E \hat{V}_j$, $\hat{A} = \hat{V}_j^T A \hat{V}_j$ and $\hat{V}_j = \mathrm{orth}\left(V_{j-1}\right)$
5:   **end if**
6:   Solve $(A + \alpha_j E)V_j = R_{j-1}$ for V_j.
7:   **if** α_j is real **then**
8:     $R_j = R_{j-1} - 2\alpha_j E V_j,\ L_j = [L_{j-1},\ V_j]$
9:   **else**
10:     $\eta_j = \sqrt{2},\ \delta_j = \frac{\mathrm{Re}\left(\alpha_j\right)}{\mathrm{Im}\left(\alpha_j\right)}$
11:     $R_{j+1} = R_{j-1} - 4\,\mathrm{Re}\left(\alpha_j\right) E(\mathrm{Re}\left(V_j\right) + \delta_j \,\mathrm{Im}\left(V_j\right))$
12:     $L_{j+1} = [L_{j-1},\ \eta_j(\mathrm{Re}\left(V_j\right) + \delta_j \,\mathrm{Im}\left(V_j\right)),\ \eta_j \sqrt{(\delta_j^2 + 1)}\ \mathrm{Im}\left(V_j\right)]$
13:     $j = j + 1$
14:   **end if**
15:   $j = j + 1$
16: **end while**
17: $D_j = -2\,\mathrm{diag}\left(\mathrm{Re}(\alpha_1), \ldots, \mathrm{Re}\left(\alpha_j\right)\right) \otimes S$

Symmetric indefinite low-rank factorization In case of the low-rank symmetric indefinite factorization, the right hand side is assumed to be of the form $W = GSG^T$ with $G \in \mathbb{R}^{n \times n_G}$, $S = S^T \in \mathbb{R}^{n_G \times n_G}$, $n_G \ll n$. Consequentially, for the one-step formulation (2.24), we have

$$
\begin{aligned}
X_j = &-2\,\mathrm{Re}\left(\alpha_j\right)(A + \alpha_j I)^{-1} GSG^T (A + \alpha_j I)^{-H} \\
&+ (A + \alpha_j I)^{-1}(A - \overline{\alpha_j} I) X_{j-1} (A - \overline{\alpha_j} I)^H (A + \alpha_j I)^{-H},
\end{aligned}
$$

that can be factored in the form

$$
\begin{aligned}
L_1 &= (A + \alpha_1 I)^{-1} G, & D_1 &= \sqrt{-2\,\mathrm{Re}(\alpha_1)} S, \\
L_j &= \left[(A + \alpha_j I)^{-1} G,\ \ (A + \alpha_j I)^{-1}(A - \overline{\alpha_j} I) L_{j-1}\right], & D_j &= \begin{bmatrix} \sqrt{-2\,\mathrm{Re}\left(\alpha_j\right)} S & \\ & D_{j-1} \end{bmatrix}.
\end{aligned}
\tag{2.29}
$$

Again, we assume $X_0 = 0$ and in addition require X to admit a factorization of the form LDL^T. Then, analogous modifications, previously applied to (2.25), for the

Algorithm 2.3 Classical low-rank Krylov subspace method

INPUT : Matrices A, $W = GG^T$, $G \in \mathbb{R}^{n \times n_G}$ defining (2.21), proper set of shift parameters $\{\mu_1, \ldots, \mu_{s-1}\} \subset \mathbb{C}$, tolerance $0 < \varepsilon \ll 1$.
OUTPUT: Z, such that $ZZ^H \approx X$.

1: Compute an orthonormal basis V of the extended or rational Krylov subspace

$$\mathcal{K}_{2s}(A, A^{-s}G) = \operatorname{span}\left\{A^{-s}G, \ldots, A^{-1}G, G, AG, \ldots, A^{s-1}G\right\} \text{ or}$$
$$\mathcal{K}_s(A, G, \mu) = \operatorname{span}\left\{G, (A - \mu_1 I)^{-1}G, \ldots, \prod_{j=1}^{s-1}(A - \mu_j I)^{-1}G\right\},$$

respectively.

2: Solve $V^TAVD + DV^TA^TV + V^TGG^TV = 0$ for $D = D^T = \tilde{Z}\tilde{Z}^T$ in factored form, using, e.g., the sign function [32, 33] or Hammerling's method [97].

3: Construct low-rank solution factor $Z = V\tilde{Z}$.

scheme (2.29), lead to the symmetric indefinite ADI procedure

$$\begin{aligned} V_1 &= (A + \alpha_1 I)^{-1}G, \qquad L_1 = V_1, \\ V_j &= V_{j-1} - (\alpha_j + \overline{\alpha_{j-1}}(A + \alpha_j I)^{-1}V_{j-1}), \\ L_j &= [L_{j-1}, V_j] \end{aligned} \tag{2.30}$$

with $D_{j_{ADI}} = -2\,\mathrm{diag}\big(\mathrm{Re}(\alpha_1), \ldots, \mathrm{Re}\big(\alpha_{j_{ADI}}\big)\big) \otimes S$, and j_{ADI} being the number of ADI iteration steps performed. This ADI formulation first appeared in [27] for Sylvester equations. The LRSIF-based equivalent of Algorithm 2.1 is given in Algorithm 2.2.

Krylov Subspace based Lyapunov Solver

The Krylov subspace methods are contained in the large class of projection based solution methods for ALEs (2.21) introduced in [167]. The main idea is to determine a low-dimensional subspace $\mathcal{V} = \operatorname{span}\{V\} \subset \mathbb{K}^n$ with $V \in \mathbb{K}^{n \times k}$, $V^TV = I_k$, $k \ll n$, and $\mathbb{K} \in \{\mathbb{R}, \mathbb{C}\}$ such that the solution to Equation (2.21) can be approximated by $X \approx VDV^T$, where $D \in \mathbb{R}^{k \times k}$ is obtained as the solution of the k-dimensional ALE

$$V^TAVD + DV^TA^TV + V^TWV = 0. \tag{2.31}$$

Since D is of dimension $k \ll n$, solution methods for small-scale dense Lyapunov equations, such as the Bartels-Stewart algorithm [13], Hammarling's method [97], the sign function method [32, 33] or any other dense Lyapunov solvers [120, 121, 122] are suitable tools for solving (2.31). The remaining task is to find an appropriate subspace to project on. In the literature the Krylov subspace methods have proven to be very effective. Most promising among these methods are the extended Krylov subspace

Algorithm 2.4 Symmetric indefinite low-rank factored Krylov subspace method

INPUT : Matrices A, $W = GSG^T$, $G \in \mathbb{R}^{n \times n_G}$, $S \in \mathbb{R}^{n_G \times n_G}$ defining (2.21), proper set of shift parameters $\{\mu_1, \ldots, \mu_{s-1}\} \subset \mathbb{C}$, tolerance $0 < \varepsilon \ll 1$.
OUTPUT: $L \in \mathbb{R}^{n \times n_L}$, $D \in \mathbb{R}^{n_L \times n_L}$ such that $LDL^T \approx X$.

1: Compute an orthonormal basis $L \in \mathbb{R}^{n \times n_L}$ of the extended or rational Krylov subspace

$$\mathcal{K}_{2s}(A, A^{-s}G) = \operatorname{span}\left\{A^{-s}G, \ldots, A^{-1}G, G, AG, \ldots, A^{s-1}G\right\} \text{ or}$$
$$\mathcal{K}_s(A, G, \mu) = \operatorname{span}\left\{G, (A - \mu_1 I)^{-1}G, \ldots, \prod_{j=1}^{s-1}(A - \mu_j I)^{-1}G\right\},$$

respectively.

2: Solve $L^T A L D + D L^T A^T L + L^T G S G^T L = 0$ for $D = D^T$.

method (EKSM) [185], also known as Krylov-Plus-Inverted-Krylov (KPIK), and the rational Krylov subspace method (RKSM) [61]. The Krylov subspace procedure for the LRCF methods based on the right hand side $W = GG^T$, is given in Algorithm 2.3. Noting that the Krylov subspace methods naturally demand for a three-term factored low-rank solution to the ALE (2.21), using $W = GSG^T$, the classical low-rank scheme can be simplified to Algorithm 2.4. In other words, exploiting the inherent structure of the Krylov subspace methods avoids the additional construction of the artificial low-rank factor Z (Algorithm 2.3, Step 3) in the classical low-rank method.

Sherman-Morrison-Woodbury Formula

From the sections above, we have seen that both, the ADI, as well as the Krylov subspace Lyapunov solvers have to perform a number of linear system solves. These shifted linear systems are in general represented by matrices of the form $F - \alpha I$. In many applications, the matrix F is given in the form $\tilde{A} + UV^T$, where U, $V \in \mathbb{R}^{n \times k}$, $k \ll n$ are dense matrices with low numerical rank. That is, even if $\tilde{A} \in \mathbb{R}^{n \times n}$ is a sparse matrix and therefore also $A = \tilde{A} - \alpha I$ is sparse, explicitly forming the product UV^T will lead to the loss of the sparsity of $A + UV^T$. Thus, for large-scale problems, the inversion of $A + UV^T$ leads to a considerable computational effort. Still, applying the Sherman-Morrison-Woodbury (SMW) formula [84] in order to invert a sparse matrix, modified by a low-rank update in the form $A + UV^T$ yields

$$\left(A + UV^T\right)^{-1} = A^{-1} - A^{-1}U(I + V^T A^{-1} U)^{-1} V^T A^{-1}. \tag{2.32}$$

That is, inverting $A + UV^T$ relies on the inversion of the sparse matrix A and the dense, but small-scale matrix $I + V^T A^{-1} U \in \mathbb{R}^{k \times k}$ that can be computed with a comparatively moderate effort.

2.1.5. Hamilton-Jacobi Theory

Following [143, Chapter 2], in this section we give the basic definitions and requirements for the solution of an optimal control problem by means of the *Hamilton-Jacobi theory*.

Consider the dynamical system

$$\begin{aligned} \dot{x} &= f(t, x(t), u(t)), \\ x(t_0) &= x_0, \end{aligned} \tag{2.33}$$

where $x \in \mathbb{R}^n$, $u \in \mathbb{R}^m$, f is a continuously differentiable function, and $x(t_0) = x_0$ defines the initial condition. In order to set up an optimal control problem for (2.33), the first step is to define an optimization criterion in terms of a performance index

$$J(x, u) = \int_{t_0}^{t_f} \ell(t, x(t), u(t))\, dt + m(t_f, x(t_f)), \tag{2.34}$$

where ℓ and m are continuously differentiable functions and $m(t_f, x(t_f))$ specifies an evaluation criterion of the final event. Note that here, we consider the finite-time interval $[t_0, t_f]$. The restriction to the finite time horizon will play an important role for the occurring equations to be solved.

The following sets and functions will be important throughout this chapter.

Definition 2.14:

a) The function $\varphi(t; t_0, x_0, u(t)) = \varphi(t; u(t))$ denotes the solution of Equation (2.33) at time t associated to the input signal $u(t)$ and initial condition x_0 at t_0.

b) $S_f \subseteq \{(t, x) : x \in \mathbb{R}^n, t \geq t_0\}$ defines the set of *admissible final events* in the sense that $(t_f, x(t_f)) \in S_f$. □

Then, the optimal control problem to be discussed reads

$$\begin{aligned} \min_u J(x, u) &= \int_{t_0}^{t_f} \ell(t, x(t), u(t))\, dt + m(t_f, x(t_f)), \\ \text{subject to} \quad & \dot{x} = f(t, x(t), u(t)), \; x(t_0) = x_0. \end{aligned} \tag{2.35}$$

In other words, the goal is to find a control u^* on $[t_0, t_f]$ that minimizes the performance index (2.34) subject to the constraints (2.33), such that $\varphi(t_f; t_0, x_0, u^*(t_f)) \in S_f$. In fact, for a linear $f(t, x(t), u(t))$ and quadratic ℓ and m, problem (2.35) defines a linear-quadratic regulator problem.

Definition 2.15 ([143, Definition 2.2]):

Let $u^*(.) \in \mathcal{U}_{ad}(t_0, t_f)$, $t_f \geq t_0$. Then, u^* is an *optimal control* related to (t_0, x_0) for the system (2.33), the performance index (2.34) and the set S_f if:

i) $\varphi(t_f; u^*(.)) \in S_f$,

ii) $J\left(\varphi(t_f, u^*(.)), u^*(.)\right) \leq J\left(\varphi(t_f, u(.)), u(.)\right)$ with $u(.) \in \mathcal{U}_{ad}(t_0, t_f)$. □

Definition 2.16 ([143, Definition 2.3]):

The function

$$H(t, x, u, \lambda) = \ell(t, x, u) + \lambda^T f(t, x, u) \tag{2.36}$$

with $\lambda \in \mathbb{R}^n$, is called the *Hamiltonian function* or short *Hamiltonian* (related to the performance index (2.34) and its constraints (2.33)). □

The quantity λ is called the *co-* or *adjoint state.* In classical optimization applications, the Hamiltonian function and the co-state correspond to the Lagrange function and the associated Lagrange multiplier.

Definition 2.17 ([143, Definition 2.4]):

The Hamiltonian is said to be regular if, as a function of u, it admits, for each x, $t \geq t_0$, λ, a unique absolute minimum $u_h^*(t, x, \lambda)$, i.e., if

$$H(t, x, u_h^*, \lambda) < H(t, x, u, \lambda), \tag{2.37}$$

$\forall u \neq u_h^*(t, x, \lambda), \forall x \in \mathbb{R}^n, \forall t \geq t_0, \forall \lambda \in \mathbb{R}^n$. □

Definition 2.18 ([143, Definition 2.5]):

Let the Hamiltonian be regular. The function u^* which verifies the inequality (2.37) is said to be the *H-minimizing control*. □

Now, a partial differential equation representing the basis of sufficient optimality conditions can be defined. It concerns a scalar function $\mathcal{V}(t, x)$ with $t \in \mathbb{R}$ and $x \in \mathbb{R}^n$.

Definition 2.19 ([143, Definition 2.6]):

Let the Hamiltonian be regular. The partial differential equation (PDE)

$$\mathcal{V}_t(t, x) + H(t, x, u_h^*(t, x, \mathcal{V}_x^T), \mathcal{V}_x^T) = 0, \tag{2.38}$$

is called *Hamilton-Jacobi equation* (HJE). □

The function $\mathcal{V}(t, x)$, satisfying the HJE (2.38) is often called the *value function*. Its partial derivatives with respect to time and space are given by $\mathcal{V}_t(t, x) = \frac{\partial \mathcal{V}(t,x)}{\partial t}$, and $\mathcal{V}_x(t, x) = \frac{\partial \mathcal{V}(t,x)}{\partial x}$, respectively.

Given the above definitions, a sufficient condition for the existence of an optimal solution to (2.35) can be stated.

Theorem 2.20 ([143, Theorem 2.1]):
Let the Hamiltonian (2.36) be regular and $u^* \in \mathcal{U}_{ad}(t_0, t_f^*)$, so that $(t_f^*, x^*(t_f^*)) \in S_f$, where $x^*(.) = \varphi(.; u^*(.))$. Let $\mathcal{V}$ be a solution of (2.38) such that:

a1) $\mathcal{V}$ is continuously differentiable;

a2) $\mathcal{V}(t, x) = m(t, x)$, $\forall (t, x) \in S_f$;

a3) $u^*(t) = u_h^*(t, x^*(t), \mathcal{V}_x(t, x)^T|_{x=x^*(t)})$, $t_0 \leq t \leq t_f^*$.

Then it follows that

t1) $u^*(.)$ is an optimal control related to (t_0, x_0);

t2) $J(t, x^*) = \mathcal{V}(t_0, x_0)$. □

The following corollary restates Theorem 2.20 such that a rule for the determination of the optimal control, based on the state equation (2.33), is given.

Corollary 2.21 ([143, Corollary 2.1]):
Let the Hamiltonian (2.36) be regular and $\mathcal{V}$ be a solution of the HJE (2.38) such that:

a1) $\mathcal{V}$ is continuously differentiable;

a2) $\mathcal{V}(t, x) = m(t, x)$, $\forall (t, x) \in S_f$.

If the equation

$$\dot{x} = f\Big(t, x(t), u_h^*\big(t, x(t), \mathcal{V}_x(t, x)^T|_{x=x(t)}\big)\Big),$$
$$x(t_0) = x_0$$

admits a solution x_c such that, for some $\tau \geq t_0$,

$$(\tau, x_c(\tau)) \in S_f,$$

then

$$u^* = u_h^*(t, x_c(t), \mathcal{V}_x(t, x)^T|_{x=x_c(t)})$$

is an optimal control related to (t_0, x_0). □

2.2. Model Order Reduction of Linear Systems

Given a partial differential equation, a spatial discretization, by e.g. the finite element method (FEM), leads to a set of ordinary differential equations. The number of degrees of freedom, i.e., the dimension of that system of ODEs, in general becomes very large.

In many applications one easily encounters systems with a dimension $\sim 10^6$. The wide field of model order reduction for linear systems is concerned with the approximation of large dynamical systems (2.1). In other words, the goal is to find a low-dimensional approximation

$$\begin{aligned} \hat{E}\dot{\hat{x}}(t) &= \hat{A}\hat{x}(t) + \hat{B}u(t), \quad \hat{x}(0) = \hat{x}_0, \\ \hat{y}(t) &= \hat{C}\hat{x}(t), \end{aligned} \tag{2.39}$$

to the original full-order model (FOM), describing the underlying physical process, where

$$\begin{aligned} \hat{E} &= W^T EV \in \mathbb{R}^{r\times r}, & \hat{A} &= W^T AV \in \mathbb{R}^{r\times r}, \\ \hat{B} &= W^T B \in \mathbb{R}^{r\times m}, & \hat{C} &= CV \in \mathbb{R}^{q\times r}. \end{aligned} \tag{2.40}$$

Further, for the reduced-order model (ROM) (2.39), we require $r \ll n$ and the output error $\|\hat{y}(t) - y(t)\|$ to be small in a suitable norm. Moreover, that means truncation matrices $V \in \mathbb{R}^{n\times r}$ and $W \in \mathbb{R}^{n\times r}$ need to be determined, such that an appropriate reduced-order model can be computed.

In the following, we give a review of the balanced truncation (BT) MOR method, as well as an insight to interpolation-based model order reduction based on Krylov subspace methods for linear time-invariant dynamical systems. An overview of the most important MOR techniques for LTI systems can, e.g., be found in [8, 78, 6, 28, 16][1]. In addition to that, in Chapter 4, certain MOR approaches for linear time-varying systems are stated.

2.2.1. Balanced Truncation

In this section, we provide the basic idea of balancing based model order reduction for LTI systems, which we will need, later on, in Section 4.1.1. These introductory statements will be extended to the LTV case in Section 4.2. The basic idea first appeared in the context of the design of digital filters in [153]. The first application in the field of model order reduction goes back to [152]. The idea of BT model order reduction is to transform the system (A, B, C) into a realization that allows to identify the states x_i, $i = 1, \ldots, n$, that are, at the same time, hard to reach and hard to observe. From an energetic point of view, those states are less important for the input-output behavior of the system and are going to be neglected in order to obtain a low-dimensional surrogate system $(\hat{A}, \hat{B}, \hat{C})$. Note that this information can be obtained from the reachability and observability system Gramians P and Q. Mathematically speaking, considering a reachable system, the expression

$$J_r = x_*^T P^{-1} x_*$$

[1] see also the MOR Wiki https://morwiki.mpi-magdeburg.mpg.de/morwiki/index.php/Main_Page

indicates the smallest amount of energy, necessary to steer the zero initial state to an arbitrary state x_*. Analogously, the observed output energy of an uncontrolled system is given by

$$J_o = x_*^T E^T Q E x_*.$$

For generalized state-space systems (E, A, B, C), with E nonsingular, the Gramians can be obtained as the solutions of the generalized algebraic Lyapunov equations

$$APE^T + EPA^T + BB^T = 0, \tag{2.41}$$

$$A^T QE + E^T QA + C^T C = 0, \tag{2.42}$$

respectively. Then, the states that are easy to control and to observe can be identified by the magnitude of the eigenvalues of the system Gramians P and $E^T QE$. Moreover, the so-called *Hankel singular values* (HSVs), given as

$$\sigma_i = \sqrt{\lambda_i(PE^T QE)},\ i = 1, \ldots, n, \tag{2.43}$$

directly reveal those states that require a large input energy to be reached and simultaneously only deliver a small amount of energy if they are observed. That is, a reduced-order model (2.39) can be obtained by truncating the HSVs of small magnitude with respect to a certain tolerance. Those HSVs can be identified using a so-called *balancing transformation* T with T nonsingular and $T^{-1}PT^{-T} = T^T E^T QET = \operatorname{diag}(\sigma_1, \ldots, \sigma_n) = \Sigma$, where $\sigma_1 \geq \cdots \geq \sigma_n$ and $T^{-1}PE^T QET = \Sigma^2$. Finding such a *balanced realization* and the computation of the truncation matrices V, W can be carried out implicitly by using, e.g., the *square-root balancing* method, see [82, 137, 197]. Assuming the system under consideration to be stable, reachable and observable, the Gramians can be obtained as the symmetric positive-definite solutions, i.e., $P = P^T > 0$, $Q = Q^T > 0$, to the reachability and observability Lyapunov equations (2.41) and (2.42), respectively. Moreover, there exist Cholesky(-like) factorizations

$$P = RR^T, \qquad Q = SS^T \tag{2.44}$$

and a balanced realization of system (2.1) can be obtained with the transformations $T = RZ\Sigma^{-\frac{1}{2}}$ and $T^{-1} = \Sigma^{-\frac{1}{2}} U^T S^T E$, where the matrices U, Σ, and Z originate from the singular value decomposition

$$S^T ER = U\Sigma Z^T = \begin{bmatrix} U_1 & U_2 \end{bmatrix} \begin{bmatrix} \Sigma_1 & \\ & \Sigma_2 \end{bmatrix} \begin{bmatrix} Z_1^T \\ Z_2^T \end{bmatrix}$$

with U_1, $Z_1 \in \mathbb{R}^{n\times r}$, $\Sigma_1 \in \mathbb{R}^{r\times r}$, U_2, Z_2, and Σ_2 of appropriate size and $U^T U = Z^T Z = I$.

Again, the matrix $\Sigma = \operatorname{diag}(\sigma_1, \ldots, \sigma_n)$ contains the HSVs of the system and the matrices V, $W \in \mathbb{R}^{n\times r}$ are obtained by truncating the $n - r$ smallest HSVs from Σ, where the σ_i, $i = 1, \ldots, n$ are ordered decreasingly and σ_{r+1} does not exceed a predefined truncation tolerance. Thus, the truncation matrices read

$$V = RZ_1\Sigma_1^{-\frac{1}{2}},\ W = SU_1\Sigma_1^{-\frac{1}{2}}. \tag{2.45}$$

Algorithm 2.5 Balanced truncation for LTI systems.

Input: $\left(E,\ A,\ B,\ C\right)$

Output: a reduced-order system $\left(\hat{E}, \hat{A}, \hat{B}, \hat{C}\right)$

1: Compute the Cholesky(-like) factors R and S of the reachability and observability Gramians $P = RR^T$ and $Q = SS^T$ satisfying the ALEs (2.41) and (2.42), respectively.

2: Compute the singular value decomposition

$$S^T ER = [\, U_1,\ U_2\,] \begin{bmatrix} \Sigma_1 & \\ & \Sigma_2 \end{bmatrix} [\, Z_1,\ Z_2\,]^T,$$

where the matrices $[\, U_1,\ U_2\,]$ and $[\, Z_1,\ Z_2\,]$ have orthonormal columns, $\Sigma_1 = \operatorname{diag}(\sigma_1, \dots, \sigma_r)$ and $\Sigma_2 = \operatorname{diag}(\sigma_{r+1}, \dots, \sigma_n)$.

3: Compute the reduced-order system (2.39) with

$$\begin{aligned} \hat{E} &= W^T EV, & \hat{A} &= W^T AV, \\ \hat{B} &= W^T B, & \hat{C} &= CV, \end{aligned} \tag{2.46}$$

where $V = RZ_1\Sigma_1^{-1/2}$ and $W = SU_1\Sigma_1^{-1/2}$.

Considering large-scale problems with $m,\ q \ll n$, the Gramians $P,\ Q$ are typically computed in terms of low-rank approximations $P \approx RR^T$ and $Q \approx SS^T$ with $R \in \mathbb{R}^{n\times k_1}$, $S \in \mathbb{R}^{n\times k_2}$, $k_1,\ k_2 \ll n$, which then serve as the Cholesky-like factors. Suitable low-rank algorithms for the computation of these factors are given in Section 2.1.4. The entire balanced truncation procedure is summarized in Algorithm 2.5.

Note that by construction $\hat{E} = I_r$. Moreover, from [6, 152] it is well known that the reduced-order system $(\hat{A}, \hat{B}, \hat{C})$ is also balanced with $\hat{P} = \hat{Q} = \operatorname{diag}(\sigma_1, \dots, \sigma_r)$ and the first r HSVs of the original system and those of the reduced-order system coincide. Thus, the following error bound [66]

$$\|\mathcal{H} - \hat{\mathcal{H}}\|_{\mathcal{H}_\infty} = \max_{\omega\in\mathbb{R}}(\|\mathcal{H}(i\omega) - \hat{\mathcal{H}}(i\omega)\|_2) \leq 2\sum_{k=r+1}^{n} \sigma_k \tag{2.47}$$

related to the full and reduced-order transfer functions $\mathcal{H},\ \hat{\mathcal{H}}$, respectively, can be stated. This allows to automatically determine the reduced order r associated to a certain truncation tolerance for the HSVs. Another favorable property of BT model order reduction is the stability preservation of the system. That is, given a stable FOM, the ROM will also be stable which is of major importance for time domain simulations based on the reduced-order model. It is noteworthy that both properties, the preservation of stability and the error bound are only proven for exact factorizations of the solutions $P,\ Q$ to the GALEs (2.41) and (2.42) and are in general not valid using the low-rank solution methods, presented in Section 2.1.4. Anyway, the influence of approximative Gramians within the square-root method has been studied in, e.g., [6, 91].

2.2.2. Interpolation-Based Model Order Reduction

In contrast to the energetic interpretation of the system behavior based on the system Gramians for the balanced truncation method, the interpolation-based reduction techniques, introduced in [207, 208, 56], aim at approximating the system related transfer function $\mathcal{H}(s)$ in the frequency domain. That means, the goal is to find a reduced-order model related to the transfer function $\hat{\mathcal{H}}(.)$ such that for a complex number $s \in \mathbb{C}$

$$\mathcal{H}(s) = \hat{\mathcal{H}}(s) \tag{2.48}$$

is satisfied. As an illustration think of a SISO system whose transfer function is a rational function in s and its degree equals the dimension of the system to be approximated. Now, an intuitive perspective of classical approximation theory is to determine a reduced-order model associated to a transfer function of lower degree interpolating the original one for a number of prescribed interpolation points lying in the complex plane. Based on the investigations in [88, 56], the following theorem states some conditions under which (2.48) can be achieved.

Theorem 2.22 (compare [41, Theorem 2.3.1]):
Consider the full-order system (E, A, B, C). Further, assume that a reduced-order system $(\hat{E}, \hat{A}, \hat{B}, \hat{C})$ is given as

$$\begin{aligned} \hat{E} &= W^T E V, & \hat{A} &= W^T A V, \\ \hat{B} &= W^T B, & \hat{C} &= C V \end{aligned}$$

with $s \in \mathbb{C} \backslash (\Lambda(A, E) \cap \Lambda(\hat{A}, \hat{E}))$ and either

i) $(sE - A)^{-1} B \in \operatorname{range}(V)$, or

ii) $(sE - A)^{-T} C^T \in \operatorname{range}(W)$.

Then it holds,

$$\mathcal{H}(s) = \hat{\mathcal{H}}(s),$$

i.e., the reduced-order transfer function $\hat{\mathcal{H}}$ is a rational matrix-valued interpolant of $\mathcal{H}$ in s. □

From the theorem above, one could easily obtain projection matrices V, W in order to obtain a suitable reduced-order system. Still, for MIMO systems the reduced-order dimension could easily exceed a certain desired bound. This drawback can be overcome by the concept of tangential interpolation, as proposed in, e.g., [79]. Here, the idea is to supplement the set of interpolation points s_i, $i = 1, \ldots, r$ with left and right tangential directions $b_i \in \mathbb{R}^m$ and $c_i \in \mathbb{R}^q$, respectively, such that the interpolation conditions

$$\mathcal{H}(s_i)b_i = \hat{\mathcal{H}}(s_i)b_i, \qquad c_i^T\mathcal{H}(s_i) = c_i^T\hat{\mathcal{H}}(s_i)$$

are satisfied at s_i along the tangential directions b_i, c_i, $i = 1, \ldots, r$. If now, in addition

$$c_i^T\mathcal{H}'(s_i)b_i = c_i^T\hat{\mathcal{H}}'(s_i)b_i \tag{2.49}$$

holds, the reduced-order transfer function $\hat{\mathcal{H}}$ is a *Hermite interpolant* of $\mathcal{H}$. Extending Theorem 2.22, the conditions (2.48) and (2.49) can be satisfied if the truncation matrices V, W, that generate the reduced-order model, associated to $\hat{\mathcal{H}}$, fulfill the subspace criteria given in the following theorem.

Theorem 2.23 ([14, Theorem 3.1]):
Let $s \in \mathbb{C}$ be such that both $sE - A$ and $s\hat{E} - \hat{A}$ are invertible. If $b \in \mathbb{C}^m$ and $c \in \mathbb{C}^q$ are fixed non-trivial vectors, then

a) if $(sE - A)^{-1}Bb \in \text{range}(V)$, then $\mathcal{H}(s)b = \hat{\mathcal{H}}(s)b$;

b) if $(sE - A)^{-T}C^Tc \in \text{range}(W)$, then $c^T\mathcal{H}(s) = c^T\hat{\mathcal{H}}(s)$;

c) if both $(sE - A)^{-1}Bb \in \text{range}(V)$ and $(sE - A)^{-T}C^Tc \in \text{range}(W)$, then $c^T\mathcal{H}'(s)b = c^T\hat{\mathcal{H}}'(s)b$. □

It is noteworthy that these criteria are immediately fulfilled constructing V, W such that

$$\begin{aligned} \text{span}\left\{(s_1E - A)^{-1}Bb_1, \ldots, (s_rE - A)^{-1}Bb_r\right\} &\subseteq \text{range}(V), \\ \text{span}\left\{(s_1E - A)^{-T}C^Tc_1, \ldots, (s_rE - A)^{-T}C^Tc_r\right\} &\subseteq \text{range}(W) \end{aligned} \tag{2.50}$$

for a given set of distinct frequencies $\{s_i\}$, left and right tangent directions $\{c_i\} \subset \mathbb{C}^q$ and $\{b_i\} \subset \mathbb{C}^m$, $i = 1, \ldots, r$, respectively. The main question remaining is the selection of interpolation sample points s_i and associated tangential directions b_i, c_i that lead to a reduced-order model with satisfactory behavior. In the literature, there is a vast variety of contributions dealing with this challenge and in addition taking into account the preservation of important system properties, as e.g., stability or passivity [6, 7, 42, 61, 88, 90, 200, 56]. The iterative rational Krylov algorithm (IRKA), introduced in [90], provides an automatic shift selection strategy and is therefore easy to apply for a wide field of users.

CHAPTER 3

OPTIMAL CONTROL AND INVERSE PROBLEMS

Contents

In this chapter, we consider a control-theoretic method for the solution of ill-posed inverse problems. To be more precise, a linear-quadratic regulator problem is formulated in order to solve the actual given inverse problem.

As an illustrative example, an inverse heat conduction problem (IHCP) is considered. In many fields of production engineering, as e.g., in chip removing processes like milling or drilling, the knowledge of process-related induced thermal loads is of great interest. Measuring the quantity of interest at the contact region of the tool and the workpiece is rather complicated or simply impossible in most settings. Therefore, the computer-aided calculation of the induced thermal loads is desired. The modeling of such processes and the associated determination of the actual induced amount of heat is a highly active research topic, see e.g. [154, 173, 174, 178] and references therein. The description of these complicated thermal processes already requires a highly accurate model. Therefore, the procedure investigated here in fact tries to avoid the sophisticated modeling of the entire production process. Instead of performing a direct simulation of the heat creation and transfer process, we solve an IHCP for the imprint temperature, denoted by u, to the workpiece from measurements $\hat{y}$ taken at accessible regions of its surface. The model equations, we consider here are given by the initial-boundary value problem

$$\begin{aligned} c_p\rho\partial_t\Theta &= \lambda\Delta\Theta && \text{in } (0,t_f)\times\Omega, \\ \lambda\partial_\nu\Theta &= \kappa^{ext}(\Theta^{ext}-\Theta) && \text{on } (0,t_f)\times\Gamma^{ext}, \\ \lambda\partial_\nu\Theta &= \kappa^{u}(u-\Theta) && \text{on } (0,t_f)\times\Gamma^{u}, \\ \Theta(0,.) &= \Theta_0, \end{aligned} \tag{3.1}$$

describing the temperature field $\Theta(t,\xi)$ at $t\in[0,T]$ and ξ being the location in the spatial domain Ω. Note that in this setting, the description of the entire physical production process can be restricted to the thermal behavior of the workpiece. The heat exchange of the domain and its exterior is modeled by the Robin-type boundary conditions on the boundary $\Gamma = \Gamma^{ext}\cup\Gamma^{u}$ with $\Gamma^{ext}\cap\Gamma^{u}=\emptyset$. Here Θ^{ext} denotes the ambient temperature, being induced at the contact boundary Γ^{ext}. Analogously, the searched for thermal input u acts on Γ^{u}. The material parameters c_p,ρ,λ, and κ^{ext} and κ^{u} denote the specific heat capacity, the density of the material, the heat conduction coefficient, and the heat transfer coefficients describing the heat exchange with the surrounding material at the boundaries Γ^{ext} and Γ^{u}, respectively. Note that in practice the determination of the heat transfer coefficients is a complicated task and requires a sophisticated transfer model.

Anyway, the resulting reconstruction problem can be formulated as:

$$\begin{gathered} \text{Find the induced load } u \text{ from given data } \hat{y} \text{ by solving} \\ Fu = \hat{y}, \end{gathered} \tag{3.2}$$

where F is a linear ill-posed operator, mapping the induced thermal load u to the measurements $\hat{y}$ taken at the surface. Solving inverse problems in general and in particular IHCPs is extensively discussed in the literature, see e.g. [65] and [39, 145] and the references therein.

3.1. Inverse Problems

In this section a brief insight to the classification of inverse problems and their most commonly used solution technique, namely the Tikhonov regularization, is given. In the literature, inverse and in particular ill-posed inverse problems first appeared in the first half of the 20th century in the fields of physics, medicine, ecology, economics and many more.

Consider an inverse problem of the form (3.2) with $u\in\mathcal{U}$, $\hat{y}\in\mathcal{Y}$ and $\mathcal{U}$, $\mathcal{Y}$ being Hilbert spaces. According to its first appearance [94], Hadamard defined the well-posedness of an inverse problem (3.2) as follows:

Definition 3.1 ([94]):

Let $F : \mathcal{U} \to \mathcal{Y}$ with Hilbert spaces $\mathcal{U}$, $\mathcal{Y}$. The inverse problem is called *well-posed* if

i) a solution $u^* \in \mathcal{U}$ of $Fu = \hat{y}$ exists for all $\hat{y} \in \mathcal{Y}$ (existence),

ii) the solution $u^* \in \mathcal{U}$ is unique (uniqueness),

iii) the solution $u^* \in \mathcal{U}$ depends continuously on the data $\hat{y} \in \mathcal{Y}$ (stability). □

Otherwise, if any of these conditions is not satisfied, the problem is called *ill-posed*. With reference to the literature, the degree of ill-posedness distinguishes mildly, moderately and severely ill-posed problems, see, e.g., [103]. In particular, according to [129], the IHCP is a severely ill-posed problem, due to its discontinuous dependence of the solutions on the data. In order to numerically overcome the ill-posedness induced by the discontinuity, the first and also most common approach, based on regularization, goes back to 1963 by Tikhonov [196]. The idea of the so-called Tikhonov-type regularization is to extend the classical approach of finding the solution u^* to the minimization of the residual norm of (3.2) in a least-squares sense

$$u^* = \operatorname{argmin} \|Fu - \hat{y}\|^2.$$

To that end, a penalization term $\alpha\|u\|^2$, in order to stabilize the inverse problem, is introduced. That is, the regularization parameter α penalizes the solutions u having large norms. Finally, one ends up with the regularized ansatz

$$\begin{aligned} u_\alpha &= \operatorname{argmin} \|Fu - \hat{y}\|^2 + \alpha\|u\|^2 \\ &= (F^*F + \alpha I)^{-1} F^* \hat{y}, \end{aligned} \tag{3.3}$$

where F^* denotes the adjoint of the linear operator F and $\alpha > 0$. It is a well known result that under reasonable conditions and $\alpha \to 0$ the family of solutions u_α converges to u^*, see, e.g., [65]. Over the years a number of extensions and more involved regularization techniques have been developed. Also other classes, such as iterative solution methods for ill-posed problems have been extensively discussed. For a detailed overview, we refer to the standard textbooks [65, 102] on inverse problems.

However, considering the solution of the inverse heat conduction problem, given in the form (3.2), originating from a PDE, relies on a full discretization with respect to space and time at the same time. Hence, the system dimension of F to be dealt with is rather large. In particular, the flexibility of adapting the time discretization according to important features of the measurement data $\hat{y}$ is given up. Therefore, in the following section we consider an optimal control related approach exploiting the given dynamics of the actual considered IHCP, described by (3.1).

3.2. Finite-Time Tracking-Type Optimal Control

In the spirit of [115, 116], we consider a linear-quadratic regulator approach to reconstruct the inaccessible temperature load induced by the physical process of interest. Well known optimal control results are used in terms of a regularization technique for ill-posed problems and at the same time exploit the dynamical structure of the heat conduction problem (3.1).

3.2.1. Model Problem and LQR Design

As a first step, using a FE space discretization, the PDE (3.1) will be rewritten into a system of ordinary differential equations in state-space form. Therefore, we consider the variational formulation

$$\begin{aligned} c_p\rho \int_\Omega \partial_t\Theta v \, d\xi &= \int_\Omega \lambda\Delta\Theta v \, d\xi \\ &= -\int_\Omega \lambda\nabla\Theta\nabla v \, d\xi + \int_\Gamma \lambda\partial_\nu\Theta v \, d\sigma \\ &= -\int_\Omega \lambda\nabla\Theta\nabla v \, d\xi + \int_{\Gamma^{ext}} \lambda\partial_\nu\Theta v \, d\sigma + \int_{\Gamma^u} \lambda\partial_\nu\Theta v \, d\sigma \\ &= -\int_\Omega \lambda\nabla\Theta\nabla v \, d\xi + \int_{\Gamma^{ext}} \kappa^{ext}(\Theta^{ext} - \Theta)v \, d\sigma + \int_{\Gamma^u} \kappa^u(u - \Theta)v \, d\sigma, \end{aligned} \tag{3.4}$$

where $v \in V = H^1(\Omega)$ define the FE test functions. Now, using a certain finite-dimensional basis $\varphi_1, \dots, \varphi_n$ the temperature field Θ and the test functions v are approximated. Further, we replace Θ^{ext} by n_{ext} discrete exterior temperatures x_k^{ext} located at boundary parts Γ_k^{ext}, $k = 1, \dots, n_{ext}$ with $\Gamma^{ext} = \bigcup_{k=1}^{n_{ext}} \Gamma_k^{ext}$ and analogously substitute u by m discrete inputs u_ℓ, acting on Γ_ℓ^u, $\Gamma^u = \bigcup_{\ell=1}^{m} \Gamma_\ell^u$, i.e., $u \leftarrow u \in \mathbb{R}^m$. That is, we allow the external temperatures x_k^{ext} to vary at different boundary parts Γ_k^{ext} combined with distinguished heat transfer coefficients κ_k^{ext}, depending on, e.g., the surface and its orientation. Similarly, different inputs u_ℓ associated to distinct transfer coefficients κ_ℓ^u are allowed at different locations of the surface of Ω. Hence, the variational form (3.4) becomes

$$
\begin{aligned}
c_p\rho \sum_{i=1}^{n} (\partial_t x)_i \int_{\Omega} \varphi_i \varphi_j \, d\xi = &- \sum_{i=1}^{n} x_i \int_{\Omega} \lambda \nabla \varphi_i \nabla \varphi_j \, d\xi \\
&- \sum_{i=1}^{n} x_i \int_{\Gamma^{ext}} \kappa^{ext} \varphi_i \varphi_j \, d\sigma - \sum_{i=1}^{n} x_i \int_{\Gamma^{u}} \kappa^{u} \varphi_i \varphi_j \, d\sigma \\
&+ \sum_{k=1}^{n_{ext}} \Theta_k^{ext} \int_{\Gamma^{ext}} \kappa_k^{ext} \varphi_j \, d\sigma + \sum_{\ell=1}^{m} u_\ell \int_{\Gamma^{u}} \kappa_\ell^{u} \varphi_j \, d\sigma, \ \forall \varphi_j, \ j = 1, \dots, n
\end{aligned}
\tag{3.5}
$$

with x_i, $i = 1, \dots, n$, being the discrete temperature values of the continuous field Θ at the FE degrees of freedom.

Then, defining the matrices

$$
\begin{aligned}
E &:= \left(c_p \rho \int_{\Omega} \varphi_i \varphi_j \, d\xi \right)_{i,j=1,\dots,n} \in \mathbb{R}^{n \times n}, \\
A &:= -\left(\lambda \int_{\Omega} \nabla \varphi_i \nabla \varphi_j \, d\xi + \kappa^{ext} \int_{\Gamma^{ext}} \varphi_i \varphi_j \, d\sigma + \kappa^{u} \int_{\Gamma^{u}} \varphi_i \varphi_j \, d\sigma \right)_{i,j=1,\dots,n} \in \mathbb{R}^{n \times n}, \\
B_\ell^u &:= \left(\kappa_\ell^u \int_{\Gamma_\ell^u} \varphi_j \, d\sigma \right)_{j=1,\dots,n} \in \mathbb{R}^{n}, \ \ell = 1, \dots, m, \\
B_k^{ext} &:= \left(\kappa_k^{ext} \int_{\Gamma_k^{ext}} \varphi_j \, d\sigma \right)_{j=1,\dots,n} \in \mathbb{R}^{n}, \ k = 1, \dots, n_{ext},
\end{aligned}
$$

Equation (3.5) can be reformulated as the generalized state-space system

$$
\begin{aligned}
E\dot{x} &= Ax + Bu + B^{ext} x_{ext}, \\
&= Ax + Bu + f
\end{aligned}
\tag{3.6}
$$

with $B = [B_1^u, \dots, B_m^u] \in \mathbb{R}^{n \times m}$, $B^{ext} = [B_1^{ext}, \dots, B_{n_{ext}}^{ext}] \in \mathbb{R}^{n \times n_{ext}}$, $x_{ext} = [x_1^{ext}, \dots, x_{n_{ext}}^{ext}]^T \in \mathbb{R}^{n_{ext}}$ and $f = B^{ext} x_{ext} \in \mathbb{R}^n$. In fact, (3.6) is an inhomogeneous generalized state-space system.

Since the goal is to reconstruct the system input u, based on data measurements $\hat{y} \in \mathbb{R}^q$ with $q \ll n$, we additionally define an output equation

$$
y = Cx, \tag{3.7}
$$

where $C \in \mathbb{R}^{q \times n}$ defines a mapping of the state variable x to the system outputs y, aiming at the reproduction of the measurements $\hat{y}$.

As a final step, an objective for the optimal control problem, covering the needs of the reconstruction purpose, is to be defined. Therefore, we consider the performance index

$$J(y,u) = \frac{1}{2} \int_{t_0}^{t_f} (y - \hat{y})^T Q (y - \hat{y}) + u^T R u \, dt + m(t_f, y(t_f)). \tag{3.8}$$

That is, minimizing (3.8) with respect to the input variable u aims at finding an optimal control u^* such that the system outputs y, defined by (3.7) and (3.6), appropriately reproduce the given measurement data $\hat{y}$. The matrices $Q \in \mathbb{R}^{q \times q}$ and $R \in \mathbb{R}^{m \times m}$ serve as weightings in the performance index (3.8) and will be described in more detail in Section 3.3. The function $m(t_f, y(t_f))$ denotes a penalization of the final output $y(t_f)$ and is defined as

$$m(t_f, y(t_f)) = \frac{1}{2} (y(t_f) - \hat{y}(t_f))^T S (y(t_f) - \hat{y}(t_f)). \tag{3.9}$$

Note that, if $Q = Q^T > 0$ and $R = R^T > 0$, the expression inside the integral of (3.8) can be written as

$$(y - \hat{y})^T Q (y - \hat{y}) + u^T R u = \|y - \hat{y}\|_Q^2 + \|u\|_R^2$$

and together with the output y defined by the output equation (3.7) and the dynamical constraints (3.6), the performance index (3.8) can be interpreted as a regularized optimization objective similar to the Tikhonov least-squares problem (3.3). The residual norm is represented by $\|y - \hat{y}\|_Q^2$ and the positive definite matrix R in a certain sense serves as the regularization parameter, penalizing the searched for control variable u. In particular, for $R = \alpha I_m$, the expression $u^T R u$ resembles a classical Tikhonov regularization. For R being diagonal with $r_{i,i} \neq r_{j,j}$, $i \neq j$ more flexibility is added to the problem. That is, different input components u_i can be regularized by different penalization parameters.

3.2.2. Solution of the Inhomogeneous Tracking Problem

The following derivation of the optimal control problem and its solution, related to the inhomogeneous system (3.6), is an extension of the classical theory for standard homogeneous state-space systems with $E = I$ that can be found in, e.g., [11, Chapter 9-9] and [143, Chapter 2].

Collecting Equations (3.8), (3.6) and (3.7), we define the finite-time LQR problem

$$\begin{aligned} \min_u J(y,u) &= \frac{1}{2}\int_{t_0}^{t_f} (y-\hat{y})^T Q (y-\hat{y}) + u^T R u \, dt + m(t_f, y(t_f)), \\ \text{subject to } E\dot{x} &= Ax + Bu + f, \\ y &= Cx. \end{aligned} \tag{3.10}$$

Provided that E is non-singular, for simplicity, we define $z := Ex$ and end up with an inhomogeneous standard state-space representation

$$\dot{z} = \tilde{A}z + Bu + f, \tag{3.11a}$$

$$y = \tilde{C}z \tag{3.11b}$$

of the given dynamics, where $\tilde{A} = AE^{-1}$ and $\tilde{C} = CE^{-1}$. Further, inserting (3.11b) into (3.8), yields

$$\begin{aligned} \min_u J(y,u) &= \frac{1}{2}\int_{t_0}^{t_f} z^T\tilde{C}^T Q\tilde{C}z - 2z^T\tilde{C}^T Q\hat{y} + \hat{y}^T Q\hat{y} + u^T R u \, dt + m(t_f, y(t_f)), \\ \dot{z} &= \tilde{A}z + Bu + f. \end{aligned} \tag{3.12}$$

According to Section 2.1.5 the Hamiltonian function $\mathcal{H}(t, z, u(t,\mu), \mu)$ of (3.12) is given by

$$\begin{aligned} \mathcal{H}(t, z, u(t,\mu), \mu) = \; & \frac{1}{2}\left(z^T\tilde{C}^T Q\tilde{C}z - 2z^T\tilde{C}^T Q\hat{y} + \hat{y}^T Q\hat{y} + u^T R u\right) \\ & + \mu^T(\tilde{A}z + Bu + f). \end{aligned} \tag{3.13}$$

Then, from the necessary and sufficient optimality conditions for (3.13), with respect to the control u, we have the first and second order derivatives

$$\mathcal{H}_u = Ru + B^T\mu \overset{!}{=} 0, \qquad \mathcal{H}_{uu} = R \overset{!}{>} 0$$

and the control u is given by

$$u(t,\mu) = -R^{-1}B^T\mu.$$

Note that the existence of R^{-1} is guaranteed by the sufficient optimality condition. Then, by Corollary 2.21 the optimal control u^* of the minimization problem (3.12) reads

$$u^*(t,\mu) = -R^{-1}B^T\mu = -R^{-1}B^T\mathcal{V}_{z^*}(t,z^*)^T = u^*\left(t, z^*(t), \mathcal{V}_z(t,z)^T|_{z=z^*(t)}\right), \tag{3.14}$$

where $z^*(t)$ is the solution trajectory of the state equation (3.11a) generated by u^*. What remains, is to find an appropriate value function, satisfying the HJE (2.38), defined in Section 2.1.5. Making the *ansatz*

$$\mathcal{V}(t,z) := \frac{1}{2}z^T X(t)z + w^T(t)z + v(t) \tag{3.15}$$

for the value function provides the derivatives

$$\mathcal{V}_t(t,z) = \frac{1}{2}z^T \dot{X}(t)z + \dot{w}^T(t)z + \dot{v}(t), \tag{3.16a}$$

$$\mathcal{V}_z(t,z) = z^T X(t) + w^T. \tag{3.16b}$$

Inserting (3.16a), (3.16b) and $u^*(t,\mathcal{V}_z^T)$ into the HJE (2.38), we obtain

$$\begin{aligned}
0 = \; & \frac{1}{2}z^T \dot{X}(t)z + \dot{w}^T(t)z + \dot{v}(t) \\
& + \frac{1}{2}\left(z^T\tilde{C}^T Q\tilde{C}z - 2z^T\tilde{C}^T Q\hat{y} + \hat{y}^T Q\hat{y} + (-R^{-1}B^T\mathcal{V}_z^T)^T R(-R^{-1}B^T\mathcal{V}_z^T)\right) \\
& + \mathcal{V}_z(\tilde{A}z + B(-R^{-1}B^T\mathcal{V}_z^T) + f) \\
= \; & \frac{1}{2}z^T \dot{X}(t)z + \dot{w}^T(t)z + \dot{v}(t) \\
& + \frac{1}{2}\left(z^T\tilde{C}^T Q\tilde{C}z - 2z^T\tilde{C}^T Q\hat{y} + \hat{y}^T Q\hat{y}\right) \\
& + \frac{1}{2}\left((-R^{-1}B^T(z^T X(t) + w^T)^T)^T R(-R^{-1}B^T(z^T X(t) + w^T)^T)\right) \\
& + (z^T X(t) + w^T)(\tilde{A}z + B(-R^{-1}B^T(z^T X(t) + w^T)^T) + f).
\end{aligned}$$

Now, a complete expansion yields

$$\begin{aligned}
0 = \; & \frac{1}{2}z^T \dot{X}(t)z + \dot{w}^T(t)z + \dot{v}(t) \\
& + \frac{1}{2}\left(z^T\tilde{C}^T Q\tilde{C}z - 2z^T\tilde{C}^T Q\hat{y} + \hat{y}^T Q\hat{y}\right) \\
& + \frac{1}{2}\left(z^T X(t)BR^{-1}B^T X(t)^T z + 2w^T BR^{-1}B^T X(t)^T z + w^T BR^{-1}B^T w\right) \\
& + z^T X(t)\tilde{A}z + w^T\tilde{A}z - z^T X(t)BR^{-1}B^T X(t)^T z - 2w^T BR^{-1}B^T X(t)^T z \\
& - w^T BR^{-1}B^T w + w^T f + z^T X(t)^T f.
\end{aligned} \tag{3.17}$$

Further, we have

$$z^T X(t)\tilde{A}z = \frac{1}{2}(z^T X(t)\tilde{A}z + z^T\tilde{A}^T X(t)^T z). \tag{3.18}$$

Using (3.18), Equation (3.17) is separated by means of the expressions quadratic in z, linear in z, and constant, respectively. This leads to the representation

$$\begin{aligned}0 =& \frac{1}{2} z^T \Big(\dot{X}(t) + \tilde{C}^T Q \tilde{C} + X(t)\tilde{A} + \tilde{A}^T X(t)^T - X(t) B R^{-1} B^T X(t)^T \Big) z \\ &+ \Big(\dot{w}^T(t) + w^T \tilde{A} - w^T B R^{-1} B^T X(t)^T - \hat{y}^T Q \tilde{C} + f^T X(t)^T \Big) z \\ &+ \dot{v}(t) - \frac{1}{2} \Big(w^T B R^{-1} B^T w - \hat{y}^T Q \hat{y} \Big) + w^T f\end{aligned}$$

of the specific HJE. That is, for

$$\begin{aligned}0 &= \dot{X}(t) + \tilde{C}^T Q \tilde{C} + X(t)\tilde{A} + \tilde{A}^T X(t)^T - X(t) B R^{-1} B^T X(t)^T, \\ 0 &= \dot{w}^T(t) + w^T \tilde{A} - w^T B R^{-1} B^T X(t)^T - \hat{y}^T Q \tilde{C} + f^T X(t)^T, \\ 0 &= \dot{v}(t) - \frac{1}{2} \Big(w^T B R^{-1} B^T w - \hat{y}^T Q \hat{y} \Big) + w^T f,\end{aligned}$$

the value function $\mathcal{V}$ from (3.15) satisfies the HJE (2.38). Hence, the optimal control given by Corollary 2.21 reads

$$u^*(t, z^*, \mathcal{V}_z^T|_{z=z^*}) = -R^{-1} B^T \mathcal{V}_{z^*}^T = -R^{-1} B^T \Big(X(t) z^*(t) + w(t) \Big), \tag{3.19}$$

where $X(t)$ is the solution of the differential Riccati equation

$$\dot{X}(t) = -\Big(\tilde{C}^T Q \tilde{C} + X(t)\tilde{A} + \tilde{A}^T X(t)^T - X(t) B R^{-1} B^T X(t)^T \Big), \tag{3.20}$$

and $w(t)$ denotes the solution to the linear inhomogeneous adjoint ODE

$$\dot{w}(t) = -\Big(\tilde{A}^T - X(t) B R^{-1} B^T \Big) w(t) + \tilde{C}^T Q \hat{y} - X(t) f. \tag{3.21}$$

Note, if $X(t)$ is a solution of (3.20) so is $X(t)^T$, and therefore, we write

$$\dot{X}(t) = -\Big(\tilde{C}^T Q \tilde{C} + X(t)\tilde{A} + \tilde{A}^T X(t) - X(t) B R^{-1} B^T X(t) \Big). \tag{3.22}$$

However, the optimal value of the performance index is

$$\mathcal{J}^*(t_0, x_0) = \frac{1}{2} z_0^T X(t_0) z_0 + w^T(t_0) z_0 + v(t_0).$$

That is, the optimal cost depends on the solutions X, w and v of the DRE (3.22), the adjoint equation (3.21) and the scalar ODE

$$\dot{v}(t) = \frac{1}{2} \Big(w^T B R^{-1} B^T w - \hat{y}^T Q \hat{y} \Big) - w^T f,$$

respectively, at time t_0.

Finding unique solutions to these ODEs requires the knowledge of an initial or final condition for each of the equations. These are given by Condition a2) of Theorem 2.20

and the penalization term of the final outputs in the performance index (3.8). To be more precise, we have

$$V(t_f, z(t_f)) = m(t_f, y(t_f))$$

and from the ansatz (3.15) for the value function and the definition (3.9) of $m(t_f, z(t_f))$, this condition yields

$$\begin{aligned}\frac{1}{2}z(t_f)^T X(t_f) z(t_f) + w^T(t_f) z(t_f) + v(t_f) &= \frac{1}{2}\big(y(t_f) - \hat{y}(t_f)\big)^T S\big(y(t_f) - \hat{y}(t_f)\big) \\ &= \frac{1}{2}z(t_f)^T \tilde{C}^T S \tilde{C} z(t_f) - \hat{y}(t_f)^T S \tilde{C} z(t_f) \\ &\quad + \frac{1}{2}\hat{y}(t_f)^T S \hat{y}(t_f).\end{aligned}$$

A simple comparison of coefficients of the left and right sides of the equality reveals the final conditions

$$X(t_f) = \tilde{C}^T S \tilde{C}, \quad w(t_f) = -\tilde{C}^T S \hat{y}(t_f), \quad v(t_f) = \frac{1}{2}\hat{y}(t_f)^T S \hat{y}(t_f).$$

Note that for the sole solution of the inverse problem under consideration, namely finding the optimal control u^*, given by Equation (3.23), the solution of v is not required.

Given all ingredients and from the re-substitution of $z = Ex$, $\tilde{A} = AE^{-1}$ and $\tilde{C} = CE^{-1}$, the solution of the optimal control problem (3.10) is given by the feedback law

$$\begin{aligned} u(t) &= -R^{-1}B^T\big(X(t)Ex(t) + w(t)\big), \\ &= -K(t)x(t) - R^{-1}B^T w(t). \end{aligned} \tag{3.23}$$

with $X(t)$ being the solution of the GDRE

$$\begin{aligned} E^T \dot{X}(t) E &= -\big(C^T Q C + E^T X(t) A + A^T X(t) E - E^T X(t) B R^{-1} B^T X(t) E\big), \\ E^T X(t_f) E &= C^T S C \end{aligned} \tag{3.24}$$

and $w(t)$ solving the generalized adjoint state equation

$$\begin{aligned} E^T \dot{w}(t) &= -\big(A^T - E^T X(t) B R^{-1} B^T\big) w(t) + C^T Q \hat{y} - E^T X(t) f, \\ E^T w(t_f) &= -C^T S \hat{y}(t_f). \end{aligned} \tag{3.25}$$

The matrix $K(t) \in \mathbb{R}^{m \times n}$ is called the gain- or feedback matrix. It is to be mentioned, that both equations, the GDRE (3.24) and the adjoint equation (3.25), need to be solved backwards in time. Comparing the expected numerical effort for the solution of the two ODEs, it turns out that the main effort in the optimal control based solution of the IHCP has to be put in solving the GDRE (3.24). Numerous DRE solvers are investigated in Chapters 5 and 6 and a detailed comparison can be found in Section 6.4.

Having a closer look at (3.23) together with the final conditions of (3.24) and (3.25), we have

$$\begin{aligned} u(t_f) &= -R^{-1}B^T\left(X(t_f)Ex(t_f) + w(t_f)\right) \\ &= -R^{-1}B^T\left(E^{-T}C^TSCx(t_f) - E^{-T}C^TS\hat{y}(t_f)\right) \\ &= -R^{-1}B^TE^{-T}C^TS\left(y(t_f) - \hat{y}(t_f)\right). \end{aligned} \tag{3.26}$$

That is, since the optimal control problem (3.10) is constructed to fit the outputs $y(t)$ to the measurements $\hat{y}(t)$ for $t \in [0, t_f)$, for stable systems and a moderate time integration step size, the outputs $y(t_f)$ at the final time $t = t_f$ will also be close to the reference outputs $\hat{y}(t_f)$. Thus, assuming y to be continuous differentiable, the difference $y(t_f) - \hat{y}(t_f)$ will by design always be close to zero, if $(y(t) - \hat{y}(t)) \approx 0$ for $t \in [0, t_f)$. This, in direct consequence, implies that for the reconstructed load, we have $u(t_f) \approx 0$. From a mathematical perspective, recall that, by the performance index, the input u is evaluated in an L_2 sense. That is, for the solution a finite number of discontinuities, representing a set of measure zero, are allowed. Thus, the single jump of the input u at time t_f does not influence the reconstruction quality in the sense of the L_2 theory. Still, the numerical procedure tries to find a smooth approximation of the searched for optimal control u and thus a smooth transition towards zero is achieved. Since the entire reconstruction is based on a limited number of measurements, given by $\hat{y} \in \mathbb{R}^q$ with $q \ll n$, the numerical results in Section 3.3 reveal an even worse behavior of u at the end of the time line of the reconstruction process. It will numerically be shown, that decreasing the amount of valuable measurement information will increase the length of the interval, where u drops down to its final condition. This also reveals that the entire reconstruction process can benefit from a clever choice of the measurement locations. Moreover, the influence of the weighting matrices, determining the final appearance of the GDRE, as well as perturbations to the underlying model are investigated by the numerical examples.

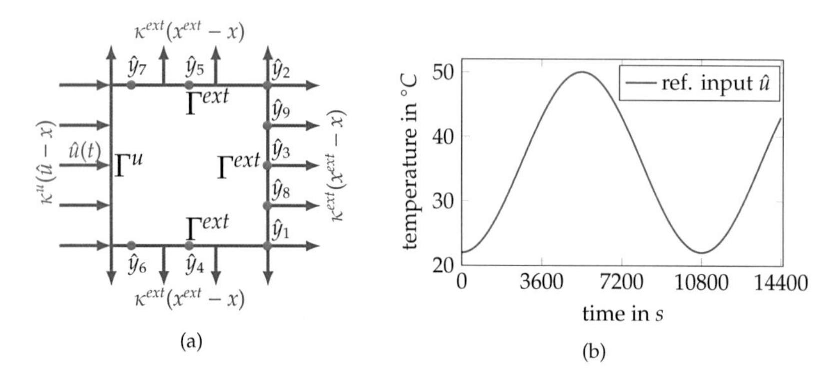

Figure 3.1.: Model domain and Robin-type boundary conditions on Γ^u (heat source) and Γ^{ext} (heat outflow) (a). Evolution of reference input $\hat{u}$ (b).

3.3. Numerical Experiments

In this section two different model problems are investigated. The first one is an artificially constructed example solely based on simulation data created by a forward simulation of the ODE system (3.6) defined on the unit square $\Omega = [0,1]^2$. The second one is a simplified real-world test example relying on a set of measurements taken at the surface of a hollow cylinder made of steel.

3.3.1. A Diffusion Problem on the Unit Square

Consider the thermal problem (3.1) defined on the unit square $\Omega = [0,1]^2$ depicted in Figure 3.1(a). The searched for scalar temperature input $u(t)$ is applied to the system at the left edge of Ω, denoted by Γ^u. The upper, right and lower edge denote the boundary Γ^{ext}, responsible for the heat exchange with the ambient. The reference data set $\hat{x}(t) \in \mathbb{R}^n$ is given by a forward simulation of the state equation (3.6) based on a FE discretization with $n = 33\,025$ degrees of freedom on the time interval $[0, 14\,400]\,s$. The reference outputs $\hat{y}(t) \in \mathbb{R}^q$, representing the measurements available for the reconstruction, are constructed by picking q temperature values at the non-input boundary Γ^{ext}, as shown in Figure 3.1(a), via the output equation (3.7). In fact, we consider a single-input-multiple-output system. In the remainder, different numbers of outputs were chosen as $q \in \{5,7,9\}$. For comparison and a better understanding, we also computed the solution to the optimal control problem (3.10) for a completely observable state, i.e., $C = I_n$ and $y = x$. The parameters $c_p = 500\frac{J}{kgK}$, $\rho = 7\,850\frac{kg}{m^3}$, $\lambda = 46\frac{W}{mK}$, and $\kappa^{ext} = 50\frac{W}{m^2K}$ of the thermal model (3.1) are chosen to belong to a workpiece consisting of heat treatable steel C45. The heat transfer coefficient describing the exchange of heat between the heat source and the domain is set to be $\kappa^u = 300 \frac{W}{m^2 K}$ for the forward simulation. The initial value x_0 is constant over the entire domain Ω and set to be 22 °C. The reference input $\hat{u} \in \mathbb{R}$ is given by the cosine-function

$$\hat{u}(t) = \frac{x_{max} - x_0}{2} \left(1 - \cos \left(\frac{2\pi t}{t_p} \right) \right) + x_0,$$

bounded by x_0 and $x_{max} = 50$ °C with a periodic time of $t_p = 10\,800\ s = 3\ h$. That is, the input reaches the maximum temperature at $t_{max} = 5\,400\ s = 1.5\ h$ the first time. Its trajectory is given in Figure 3.1(b).

The reconstruction of the reference input depends on several properties of the optimal control problem. In the following, the results based on the numbers of measurements and the choice of the weighting matrices Q, R and S are presented. Basically, the weighting matrices are chosen as multiples of the identity. The matrix Q penalizes deviations of the outputs $y(t)$ to the measurements $\hat{y}(t)$ for $t \in [0, t_f)$, the matrix R regulates the costs of the input and matrix S is used for penalization of the outputs at $t = t_f$. Since we aim at reconstructing the system input u, we choose $R = I_m$ in order to introduce the pure input portion to the performance index (3.8) of the problem. In practice, the exact determination of the heat transfer coefficients is a very challenging task. As mentioned at the beginning of this chapter, in particular the description of the entire machining process and thus of the boundary condition at the input boundary, is a highly active research field. Therefore, we also investigate the solution of the LQR-type optimal control problem (3.10) with respect to a perturbed heat transfer coefficient κ^u at the input boundary Γ^u.

Number of Outputs

Since the reconstruction of the induced temperature to the system is based on the measurement data taken at the non-input boundaries, here we investigate the influence of the density of given information in terms of the number of measurements taken. Therefore, at first we choose a pure weighting of the outputs via the constant matrices $Q = 5e3 \cdot I_q$ and $S = I_q$. Figure 3.2 shows the reconstructed input based on $q = 5,\ 7,\ 9$ and $q = n$ measurements. The case $q = n$ is in practice and in particular for large-scale problems not practicable. In order to keep the computational effort acceptable for this comparison an FE grid of $n = 81$ degrees of freedom was chosen. From Figure 3.2(a) it can be seen that the reference solution is reconstructed up to a relative error in the range $1e$-3 up to $1e$-4. Having a closer look to the end of the time interval (Figure 3.2(b)), we observe the reconstructed solution to decay towards zero. This drop is due to the terminal conditions for the GDRE (3.24) and the adjoint equation (3.25) and the resulting relation (3.26). Still, it can be seen that increasing the given information by increasing the number of outputs, the time at which the drop occurs can be moved towards the final time. In particular, for $q = n$, we observe the

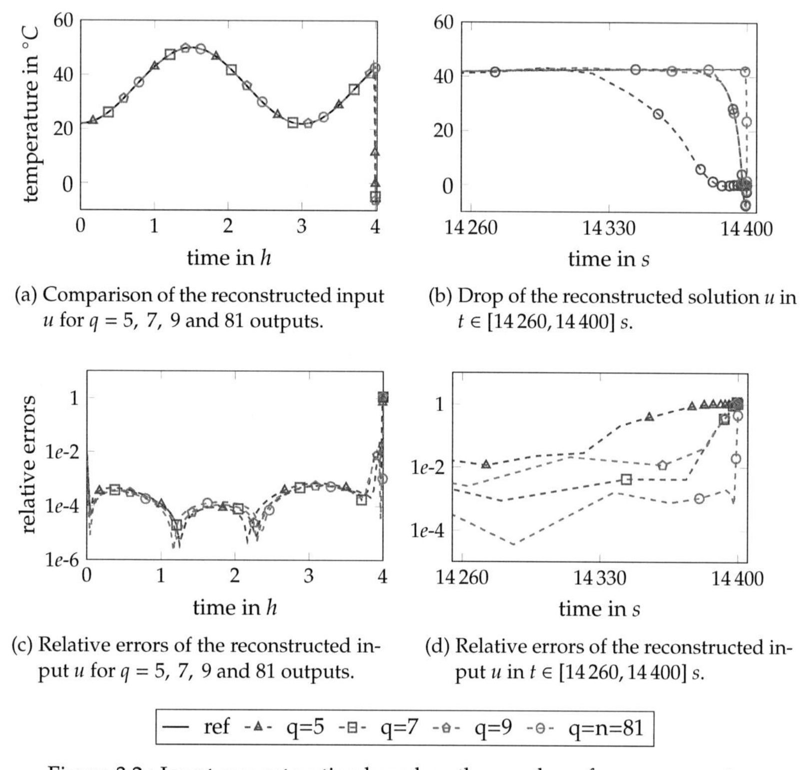

(a) Comparison of the reconstructed input u for $q = 5,\ 7,\ 9$ and 81 outputs.

(b) Drop of the reconstructed solution u in $t \in [14\,260, 14\,400]$ s.

(c) Relative errors of the reconstructed input u for $q = 5,\ 7,\ 9$ and 81 outputs.

(d) Relative errors of the reconstructed input u in $t \in [14\,260, 14\,400]$ s.

Figure 3.2.: Input reconstruction based on the number of measurements q.

smooth decay to become a jump. Figures 3.2(c) and 3.2(d) show the corresponding relative errors to Figures 3.2(a) and 3.2(b), respectively. The results also reveal that including measurements $\hat{y}_8, \hat{y}_9$ does not mentionable affect the reconstruction accuracy. This is not too surprising since the measurements $\hat{y}_j$, $j = 1, 2, 3$ already include the information contained in the right edge of the unitsquare. Basically, the entire reconstruction procedure benefits from a clever choice of the measurement locations such that the essential dynamical behavior is captured.

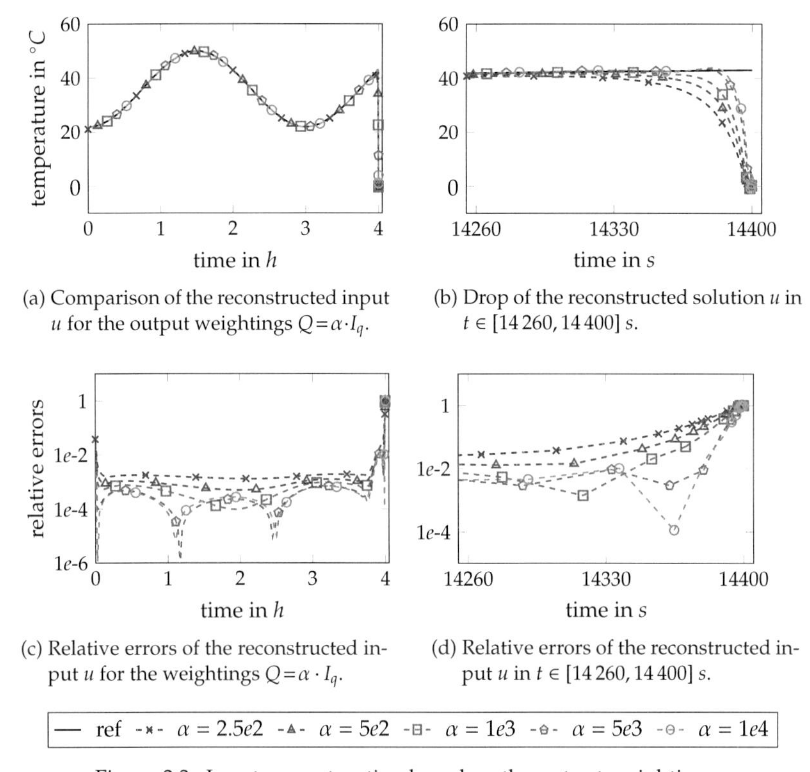

(a) Comparison of the reconstructed input u for the output weightings $Q=\alpha \cdot I_q$.

(b) Drop of the reconstructed solution u in $t \in [14\,260, 14\,400]$ s.

(c) Relative errors of the reconstructed input u for the weightings $Q=\alpha \cdot I_q$.

(d) Relative errors of the reconstructed input u in $t \in [14\,260, 14\,400]$ s.

Figure 3.3.: Input reconstruction based on the output weighting α.

Weighting Matrices

Here, a FE discretization of $n = 289$ degrees of freedom is chosen. The representation is restricted to the results for $q = 7$ measurements. According to the above example and Equation (3.26), the weighting of the outputs at the final time t_f, i.e., matrix S in the final conditions for the DRE and the adjoint equation does not have a notable influence to the computations and therefore is again set to be $S = I_q$. Then, Figure 3.3 presents the influence of the output weighting Q to the solution of the reconstruction problem. For the computations we chose $Q = \alpha I_q$ and investigate the solution for a changing scalar value α. As expected, the accuracy of the reconstructed input u compared to the reference temperature source improves for an increasing penalization of the output difference $y - \hat{y}$ in the performance index (3.8). Still, this holds only up to a certain limit. Figure 3.3(b) reveals that for $\alpha > 5e3$ the control u starts to oversteer.

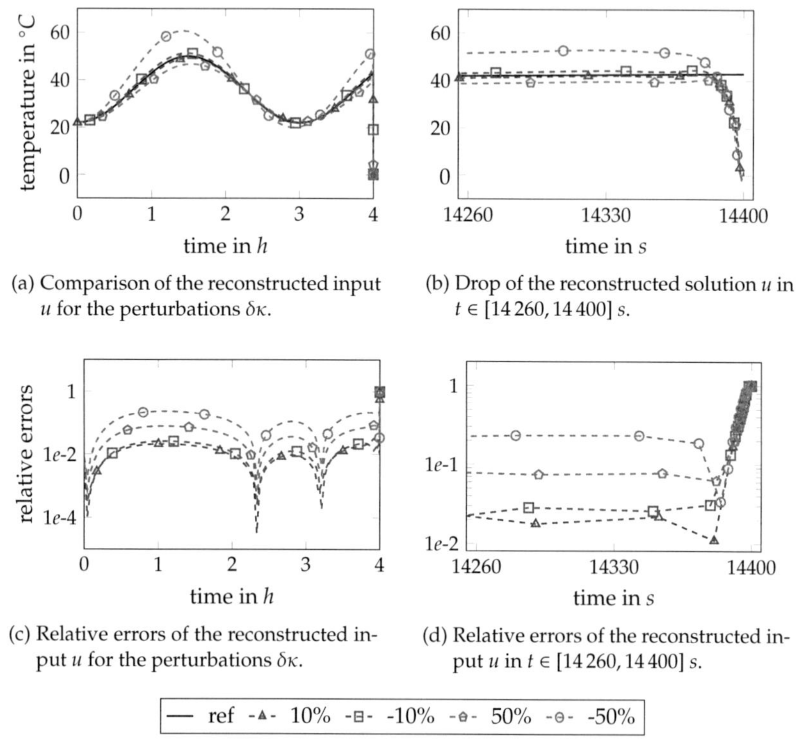

(a) Comparison of the reconstructed input u for the perturbations $\delta\kappa$.

(b) Drop of the reconstructed solution u in $t \in [14\,260, 14\,400]$ s.

(c) Relative errors of the reconstructed input u for the perturbations $\delta\kappa$.

(d) Relative errors of the reconstructed input u in $t \in [14\,260, 14\,400]$ s.

Figure 3.4.: Perturbed heat transfer coefficient κ^u at input boundary Γ_1.

Perturbed heat Transfer Coefficient

In general the reference solution of the optimal control problem (3.10) does not exist and the error of the solution cannot be computed. As mentioned before the result of the reconstruction problem is only reliable if a sufficiently accurate model was used for the tracking procedure. Assuming that any information at the input boundary is missing, it seems to be obvious to investigate, e.g., the influence of a perturbation to the model, describing the PDE. Here, a noisy heat transfer coefficient κ^u at the input boundary Γ^u is considered. In Figure 3.4 the reconstructed inputs according to perturbations $\delta\kappa$ of 10 %, -10 %, 50 % and -50 % are depicted. One observes that a perturbation of $\delta\kappa = \pm 10$ % to the heat transfer coefficient still yields a relative reconstruction error in a single-digit percent range.

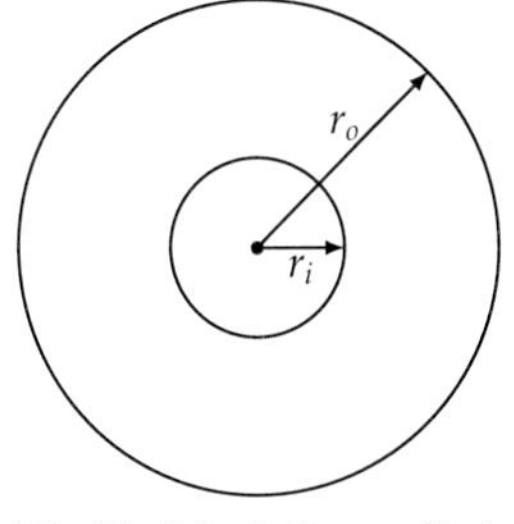

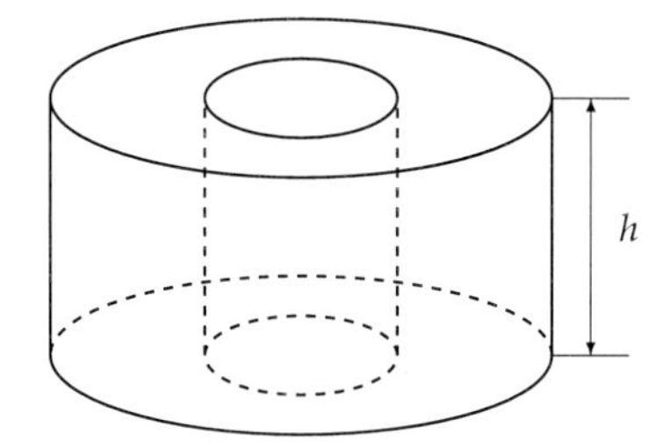

(a) Radii of the hollow cylinder.

(b) Height of the hollow cylinder.

Figure 3.5.: Model domain Ω

3.3.2. Real World Hollow Cylinder

For the second example, we consider the 3-dimensional steel hollow cylinder depicted in Figure 3.5. The domain represents an experimental workpiece with an inner radius $r_i = 47.75\ mm$, an outer radius $r_o = 113\ mm$ and a height of $h = 102\ mm$. The inner cylindrical boundary serves as the input boundary Γ^u. The remaining boundaries denote Γ^{ext}. In this example, the boundary Γ^{ext} is divided into the lower surface, which we denote as Γ^{ext}_{low}, and the upper and outer boundary parts are collectively denoted by Γ^{ext}_{out}. The boundary Γ^{ext}_{low} is set to be isolated and Γ^{ext}_{out} inherits the Robin boundary condition from Γ^{ext}. That is, the model equation (3.1) becomes

$$\begin{aligned} c_p\rho\partial_t\Theta &= \lambda\Delta\Theta && \text{in } (0,t_f)\times\Omega, \\ \lambda\partial_\nu\Theta &= 0 && \text{on } (0,t_f)\times\Gamma^{ext}_{low}, \\ \lambda\partial_\nu\Theta &= \kappa^{ext}(\Theta^{ext}-\Theta) && \text{on } (0,t_f)\times\Gamma^{ext}_{out}, \\ \lambda\partial_\nu\Theta &= \kappa^{u}(u-\Theta) && \text{on } (0,t_f)\times\Gamma^{u}, \\ \Theta(0,.) &= \Theta_0. \end{aligned} \tag{3.27}$$

The model problem, we consider here, is intended to simulate a heavily simplified drilling process. The thermal input u, representing the friction driven thermal load of the process, is induced by a set of heating mats at the inner boundary Γ^u as depicted in Figure 3.6(a). As in practical applications, for this example, the exact heating behavior of the heating devices is unknown and therefore there is no reference input available. The measurements, being the basic information of the reconstruction procedure, are taken by 6 sensors, located at the isolated bottom (boundary Γ^{ext}_{low}) of the annulus as presented in Figure 3.6(b). These sensors are numbered with 1-6 from the inner to the outer cylinder. The measurements are taken over the time interval $t \in [0, 55\,320]\ s$. For illustration, the given data in Sensors 1, 4 and 6 is presented in Figure 3.7(a). For the reconstruction of the temperature input, induced by the heat mats, we consider a FE model of dimension $n = 11\,318$. The weighting matrices Q, R, S in the performance index (2.34) are, as in the previous example, chosen to be multiples of the identity

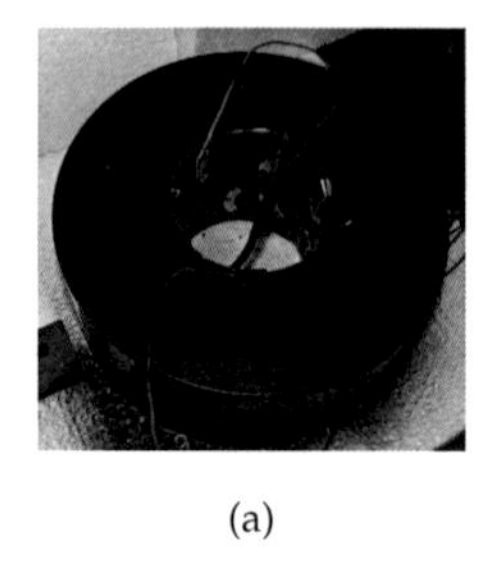

(a)

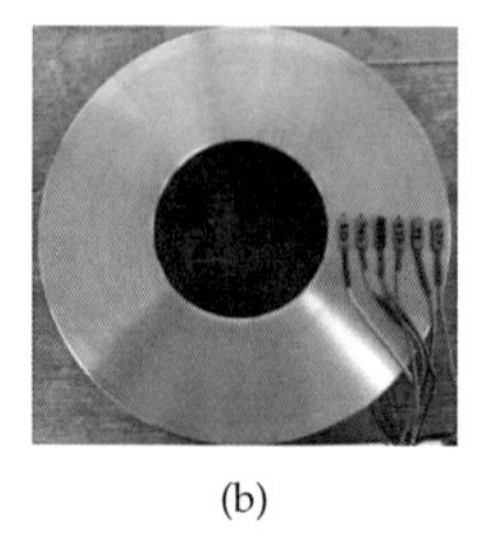

(b)

Figure 3.6.: Heating mats at the inner boundary, attached by magnets (a). Sensors at the bottom boundary Γ_{low}^{ext} (b).

of appropriate size. Note, the choice of the weightings can be further improved by using more advanced information of e.g., a weighted importance of the measurement nodes. Here, $Q = 2\,500 \cdot I_q, R = I_m$ and $S = 100 \cdot I_q$. The scaling factors of the several identities are empirically determined in order to ensure a moderate error of the computed and the desired output (measurements). Since the computational effort of computing the solution of the GDRE (3.24) of dimension $n = 11\,318$ over the time interval $t \in [0, 55\,320]$ s is way too expensive using the currently available solvers, we use balanced truncation model order reduction in order to reduce the number of relevant degrees of freedom in advance of solving the optimal control problem (3.10) by solving the associated GDRE. Finally, as a low-dimensional surrogate model of the state-space representation of (3.27), similar to (3.6), we consider the reduced-order system

$$\begin{aligned} \hat{E}\dot{\hat{x}} &= \hat{A}\hat{x} + \hat{B}u + \hat{f}, \\ \hat{y} &= \hat{C}\hat{x} \end{aligned} \tag{3.28}$$

of dimension $r = 25$. Here, the reduced dimension was empirically chosen. Computing the solution of the optimal control problem with respect to (3.28), we obtain the reconstructed heat source u presented in Figure 3.7(b). Since we cannot compare the result to a reference solution, here we compare the measurements y and the computed outputs $\hat{y}$ induced by the reconstructed input u. The results are depicted in Figure 3.8(a) and the corresponding relative errors are given in Figure 3.8(b). The achieved errors for the outputs lie within the per mille range such that the result for the reconstructed input u seems to be reliable. Still, note that the output error is not a reliable criterion for correctness of the reconstructed input. That is, given a perturbed model (3.6), the optimal control problem may result in a certain solution u reproducing the measurements up to a desired error tolerance but will be based on violated model data. This in particular means that a highly accurate understanding of the thermal process and its thorough modeling is absolutely essential for the entire reconstruction procedure. The oscillations in u arise from the noisy measurements.

These uncertainties can be reduced by filtering the given measurements or using a linear quadratic Gaussian (LQG) design [143] instead of the described LQR ansatz. For completeness, note that the latter will lead to the solution of an additional, so-called *filter differential Riccati equation* (FDRE) and thus, without taking special structure of both, the GDRE and FDRE, into account, roughly doubles the computation time.

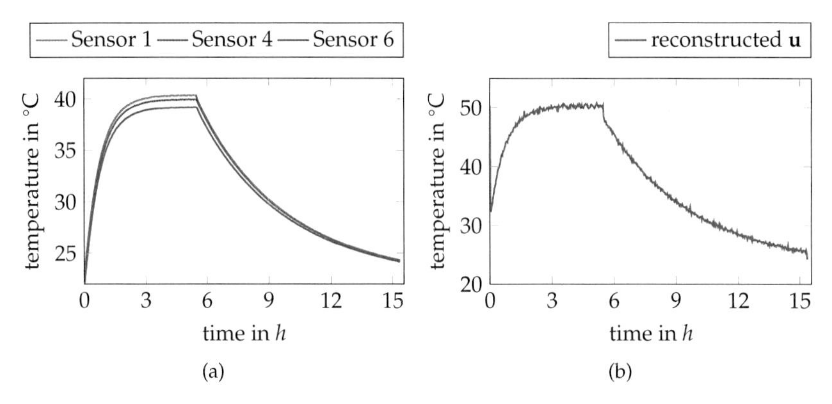

Figure 3.7.: Measured data y in Sensors 1, 4, 6 (a) and the corresponding reconstructed input u (b).

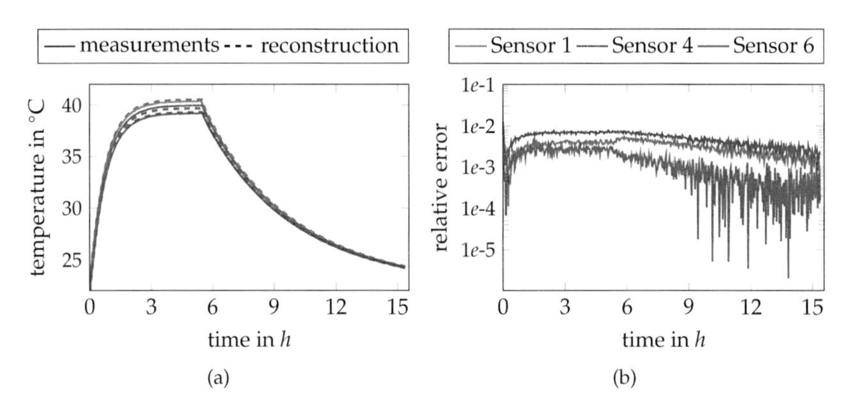

Figure 3.8.: Comparison of the measured outputs y and the computed outputs $\hat{y}$ (a) and the associated relative errors for Sensors 1 ,4 and 6.

3.4. Summary and Conclusion

An IHCP problem regarding the reconstruction of a thermal load based on given measurement data is considered. For reasons of comparison a very brief insight into ill-posed inverse problems is given, followed by a classical solution approach in terms of Tikhonov regularization. Based on the example of a specific IHCP, the inverse problem is reformulated as an optimal control problem of tracking-type, capable of exploiting the inherent dynamics of the heat equation. Therefore, the given PDE is semi-discretized and transferred into a control-theoretically motivated state-space representation. Based on the resulting inhomogeneous generalized state-space system (3.6), the Hamilton-Jacobi theory, reviewed in Section 2.1.5, is applied to the tracking-type optimal control problem. It is stated that the main ingredient for the computation of the quantity of interest, is the solution of a GDRE. Further, it is shown that the optimal control $u(t_f)$, representing the thermal load to be reconstructed at the final time t_f, will always be zero or at least close to zero by construction. Still, this is not a crucial problem. Since, the final condition of the input u at the single time instance t_f represents a set of measure zero, the single discontinuity, given by the theoretical design of the problem, does not influence the reconstruction quality in the $\mathcal{L}_2$ sense. However, two numerical examples show the performance of the presented procedure, whereas a critical evaluation regarding the reliability of the results is stated.

CHAPTER 4

MODEL ORDER REDUCTION OF LINEAR TIME-VARYING SYSTEMS

Contents

Adapting the MOR idea for LTI systems, from Section 2.2, to LTV systems

$$\begin{aligned} E(t)\dot{x}(t) &= A(t)x(t) + B(t)u(t),\ x(t_0) = x_0, \\ y(t) &= C(t)x(t) \end{aligned} \tag{4.1}$$

yields a reduced-order model of the form

$$\begin{aligned} \hat{E}(t)\dot{\hat{x}}(t) &= \hat{A}(t)\hat{x}(t) + \hat{B}(t)u(t), \quad \hat{x}(t_0) = \hat{x}_0, \\ \hat{y}(t) &= \hat{C}(t)\hat{x}(t) \end{aligned} \tag{4.2}$$

with $x(t) \approx V(t)\hat{x}(t)$ and the low-dimensional, time-dependent matrices

$$\begin{aligned} \hat{E}(t) &= W(t)^T E(t)V(t) \in \mathbb{R}^{r(t)\times r(t)}, & \hat{A}(t) &= W(t)^T \left(A(t)V(t) - E(t)\dot{V}(t)\right) \in \mathbb{R}^{r(t)\times r(t)}, \\ \hat{B}(t) &= W(t)^T B(t) \in \mathbb{R}^{r(t)\times m}, & \hat{C}(t) &= C(t)V(t) \in \mathbb{R}^{q\times r(t)}. \end{aligned} \tag{4.3}$$

Model order reduction for LTV systems (4.1) is a highly active research field. In that context, main effort has been put into the extension of the BT theory for LTI models to LTV systems, see, e.g., some early references [183, 202]. In the last two decades, BT for LTV systems received increasing attention. A number of different error bounds are presented in, e.g., [144, 128, 171, 68], whereas in [50, 51, 201] numerical issues and algorithms are discussed. However, the main challenge in MOR for LTV systems is the efficient computation and storage of the time-dependent truncation matrices $V(t)$, $W(t)$. Moreover, the time-dependence of the projection basis calls for an online application, i.e., the reduced-order matrices (4.3) need to be computed for every single time instance. Thus, a clever strategy for the computation of global or at least locally piecewise constant truncation matrices is of particular interest. In the literature there is a number of contributions, transforming the LTV system into LTI or even linear parameter-varying (LPV) systems. In [105, 161] the MOR approach is converting the LTV system into an LTI system using a time discretization and applying a recycled Krylov subspace method. For systems with time-dependent input and output matrices, and constant dynamics, i.e., E, A constant, in [195] a two-step MOR approach is presented. Therein the time-dependence of the input and output matrices is isolated and included in the input $u(t)$ or output $y(t)$, respectively. The resulting auxiliary system is of LTI structure and therefore the classical MOR techniques for LTI systems can be applied to the system. Interpreting the time-dependence as a special case of a parameter-dependence leads to parametric model order reduction approaches for LPV systems. In that context, in [74, 75] PMOR based on matrix interpolation [4, 155] was applied to moving load problems. An adaption of the PMOR scheme based on matrix interpolation to systems with a parameter, explicitly depending on time, can be found in, e.g., [53].

In this thesis, three different model order reduction approaches for the approximation of LTV systems are considered. In Section 4.1.1 a procedure based on a piecewise constant approximation of the global LTV system in terms of a switched linear system (SLS), consisting of several LTI subsystems, is presented. Then, in Section 4.1.2 a parametric model order reduction scheme, using interpolatory projection methods [14] is applied to an LPV approximation of the underlying LTV system. Finally, in Section 4.2 we review a numerical algorithm for the BT method applied to the original LTV system. Further, we discuss its applicability to large-scale LTV models with regard to the solution of the arising differential Lyapunov equations and the efficient computation of the time-dependent reduced-order matrices (4.3). In Section 4.3 the performance of the methods is presented.

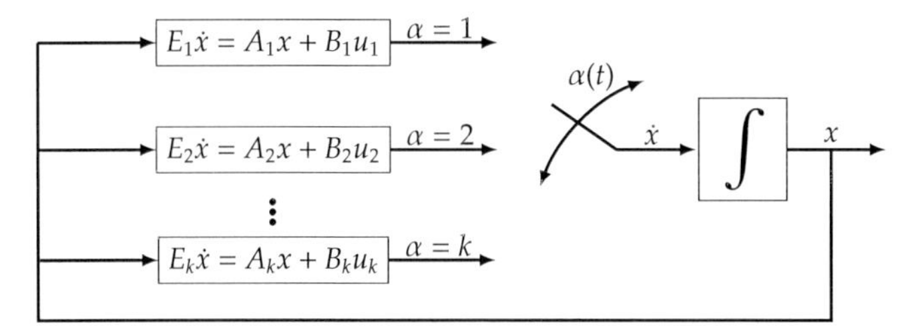

Figure 4.1.: Time-dependent switching of linear time-invariant systems.

4.1. Model Order Reduction using LTI Model Approximations

4.1.1. MOR for Switched Linear Systems

The first approach, approximating the time-dependence of a linear system (4.1), relies on the reformulation of the actual LTV system into a finite number $k \in \mathbb{N}$ of linear dynamical subsystems. These local subsystems are assumed to be of LTI structure. Together with the locally piecewise constant function α, responsible for the selection process, we arrive at the auxiliary system

$$\begin{aligned} E_\alpha \dot{x} &= A_\alpha x + B_\alpha u, \\ y &= C_\alpha x. \end{aligned} \tag{4.4}$$

In fact, system (4.4) is a SLS [141] and α denotes the switching signal, taking its values from the index set $\mathfrak{J} = \{1, \ldots, k\}$. Regarding the investigations in [206], switching signals can be classified into two categories. The first of which is the time-dependent switching, where the switching signal solely depends on time t. The second form is the state and/or output-dependent switching, where the switching is intrinsically driven by the state/output trajectory. In general, switched systems are a subclass of hybrid dynamical systems [199], consisting of a family of continuous (and/or discrete) time subsystems and a rule that performs the switching between the systems. However, in the remainder, we restrict to the case, where all subsystems are continuous-time LTI systems connected by time-dependent switching with

$$\alpha(t) : t \mapsto \{1, \ldots, k\} = \mathfrak{J}. \tag{4.5}$$

A sketch of time-dependent switching between LTI systems is given in Figure 4.1. Moreover, the intervals, where specific subsystems remain active, may be arbitrary small, but an infinite number of switches within finite time is excluded [141]. Concerning stability of switched systems it is noteworthy that arbitrary switching may result in an unstable switched linear system, even if every single subsystem is asymptotically stable, see, e.g, [141]. The stability theory for switched systems is well understood and for switched linear systems with an arbitrary switching signal the following theorem holds.

Theorem 4.1 ([141, Section 2]):
A switched linear system is asymptotically stable under arbitrary switching if and only if the subsystems share a common quadratic Lyapunov function (CQLF). □

For SLS it is natural to consider a common quadratic Lyapunov function of the form

$$\mathcal{L}(x) = x^T P x, \tag{4.6}$$

where P is a symmetric positive definite solution of the linear matrix inequality

$$A_\alpha P + P A_\alpha^T \leq -Q, \quad \alpha \in \mathfrak{J},$$

for some symmetric positive definite matrix Q.

However, switched linear systems are often applied for the approximation of nonlinear systems. This is an appropriate procedure for nonlinear systems dealing with sudden property changes, as, e.g., vibration or starting simulations of motors. One of the most important advantages is that the use of SLS allows the application of the well studied control related theory for linear systems such as model order reduction. This, of course, also holds for the approximation of an LTV system (4.1) in terms of a family of LTI subsystems. So far, two MOR strategies have been considered in the literature. The first approach aims at finding a reduced-order model for the overall SLS. In [179, 180, 181] a balanced truncation based method using generalized Gramians has been proposed. Based on a common projection, this approach guarantees the stability of the resulting reduced-order model. Still, this comes at the cost of solving systems of linear matrix inequalities (LMIs) instead of the Lyapunov equations in (2.41) and (2.42) for the classical balanced truncation framework (see Section 2.2.1), and further is at the expense of the attainable accuracy for the several subsystems. That is, in terms of accuracy the common projection framework cannot compete with separately reduced subsystems. Therefore, the second strategy is built on local reduction in order to find an accurate reduced-order approximation for each subsystem. For that, any projection-based model order reduction technique, such as Krylov subspace methods or BT, can be applied. In contrast to the first strategy, the direct application of the second one does not guarantee the stability of the reduced-order model, even if the single low-dimensional subsystems are stable. That is, the preservation of stability of the overall SLS requires additional effort. A balanced truncation based procedure, using an averaging strategy of simultaneously balanced Gramians of the subsystems is proposed in [151]. But, this approach is rather theoretical since it relies on the commutativity of the Gramians of a single subsystem and in addition requires the Gramians P_i, Q_i and P_j, Q_j to satisfy $P_i Q_j = P_j Q_i$ for $i, j \in \mathfrak{J}$. In practice, and in general for large-scale problems, these conditions are difficult to verify. That makes the procedure inapplicable.

However, note that reducing every subsystem individually, in general leads to reduced-order subsystems

$$\begin{aligned} \hat{E}_\alpha \dot{\hat{x}}_\alpha &= \hat{A}_\alpha \hat{x}_\alpha + \hat{B}_\alpha u, \\ \hat{y}_\alpha &= \hat{C}_\alpha \hat{x}_\alpha \end{aligned} \tag{4.7}$$

with $\hat{E}_\alpha$, $\hat{A}_\alpha \in \mathbb{R}^{r_\alpha \times r_\alpha}$, $\hat{B}_\alpha \in \mathbb{R}^{r_\alpha \times m}$, and $\hat{C}_\alpha \in \mathbb{R}^{q \times r_\alpha}$ and reduced states $\hat{x}_\alpha \in \mathbb{R}^{r_\alpha}$, $x \approx V_\alpha \hat{x}_\alpha$. Since the states $\hat{x}_\alpha$ may live in different subspaces, the subsystems cannot have a common Lyapunov function of the form (4.6), even if the matrices P_α are equal. Still, stability of the reduced-order SLS can be obtained. In [81] a stability-preserving MOR approach, using individual subspaces, is proposed. For this, we define the Lyapunov function

$$\mathcal{L}(\hat{x}_\alpha(t)) := \begin{cases} \mathcal{L}_1(\hat{x}_1(t)) & \text{if } \alpha = 1, \\ \quad\vdots & \\ \mathcal{L}_k(\hat{x}_k(t)) & \text{if } \alpha = k, \end{cases} \tag{4.8}$$

of the reduced-order SLS. That is, the Lyapunov function (4.8) of the overall reduced SLS (4.7) takes its value according to the Lyapunov function of the active subsystem. Then, the following theorem can be stated.

Theorem 4.2 ([81, Theorem 3]):
Given a SLS with asymptotically stable subsystems. Then, a reduced switched linear system with individually reduced-order subsystems is asymptotically stable if the associated Lyapunov function $\mathcal{L}(\hat{x}_\alpha(t))$ decreases monotonically, i.e:

1. The reduced subsystems α have individual Lyapunov functions, satisfying

$$\mathcal{L}_\alpha(\hat{x}_\alpha(t)) > 0 \text{ and } \dot{\mathcal{L}}_\alpha(\hat{x}_\alpha(t)) < 0$$

   and therefore are asymptotically stable.

2. Switching from subsystem $\alpha = i$ to $\alpha = j$ at time t_s, the value of the individual Lyapunov functions, which may be discontinuous, does not increase:

$$\mathcal{L}_j(\hat{x}_j(t_s)) \leq \mathcal{L}_i(\hat{x}_i(t_s)). \qquad \square$$

Thus, based on Theorem 4.2, the stability-preserving procedure basically consists of two main steps. At first, stable low-dimensional approximations of the form (4.7) to the given stable full-order subsystems (4.4) need to be computed. Using, e.g., the balanced truncation model order reduction the stability of the subsystems is preserved and in addition the error bound (2.47) for each of the subsystems is provided, see Section 2.2.1. The second step is to ensure, that the Lyapunov function (4.8), associated to the reduced-order SLS is monotonically decreasing. In particular, this means it does not increase at switching times t_s. This can be satisfied applying certain state transformations to the reduced states $\hat{x}_\alpha$. Therefore, assume that the transformed reduced subsystems, obtained by a state transformation $\tilde{x}_\alpha = T_\alpha \hat{x}_\alpha$ share the Lyapunov matrix $\tilde{P}_\alpha = I$. From the stability theory by Lyapunov, the monotonically decrease of the norm $\|\tilde{x}_\alpha(t)\|$ is guaranteed, see, e.g., [112, 113], and therefore the Lyapunov function $\mathcal{L}(\hat{x}_\alpha(t)) = \hat{x}_\alpha^T \hat{x}_\alpha = \|\hat{x}_\alpha\|^2$ also decreases monotonically. Further, we obtain the transformed low-dimensional subsystems

$$\begin{aligned} \tilde{E}_\alpha \dot{\tilde{x}}_\alpha &= \tilde{A}_\alpha \tilde{x}_\alpha + \tilde{B}_\alpha u, \\ \tilde{y}_\alpha &= \tilde{C}_\alpha \tilde{x}_\alpha \end{aligned} \tag{4.9}$$

with

$$\begin{aligned} \tilde{E}_\alpha &:= T_\alpha \hat{E}_\alpha T_\alpha^{-1}, & \tilde{A}_\alpha &:= T_\alpha \hat{A}_\alpha T_\alpha^{-1} \\ \tilde{B}_\alpha &:= T_\alpha \hat{B}_\alpha, & \tilde{C}_\alpha &:= \hat{C}_\alpha T_\alpha^{-1}. \end{aligned}$$

Given stable reduced-order subsystems (4.7), the systems (4.9) remain stable by the similarity transformation T_α. The question remaining is the choice of the state transformation T_α. From the property $\tilde{P}_\alpha = I$, being the common Lyapunov matrix of the systems (4.9), we have

$$\begin{aligned} 0 &> \tilde{E}_\alpha^T I \tilde{A}_\alpha + \tilde{A}_\alpha^T I \tilde{E}_\alpha \\ &= (T_\alpha \hat{E}_\alpha T_\alpha^{-1})^T I (T_\alpha \hat{A}_\alpha T_\alpha^{-1}) + (T_\alpha \hat{A}_\alpha T_\alpha^{-1})^T I (T_\alpha \hat{E}_\alpha T_\alpha^{-1}). \end{aligned} \tag{4.10}$$

Then, defining the decomposition $\hat{P}_\alpha := T_\alpha^T T_\alpha$ with $\hat{P}_\alpha > 0$, Equation (4.10) can be written in the form

$$\hat{E}_\alpha^T \hat{P}_\alpha \hat{A}_\alpha + \hat{A}_\alpha^T \hat{P}_\alpha \hat{E}_\alpha < 0. \tag{4.11}$$

In order to find a solution $\hat{P}_\alpha > 0$ of the Lyapunov inequalities (4.11), one can choose any $Q > 0$ and solve the ALE

$$\hat{E}_\alpha^T \hat{P}_\alpha \hat{A}_\alpha + \hat{A}_\alpha^T \hat{P}_\alpha \hat{E}_\alpha = -Q \tag{4.12}$$

instead, see [40, Section 2.7.1]. Note that this parametrizes the entire set of solutions to the LMIs (4.11) by the choice of Q. Given the solutions $\hat{P}_\alpha > 0$, the factors T_α can, e.g., be obtained by a Cholesky factorization. Moreover, since the ALEs (4.12) are of low dimension, we can use small-scale Lyapunov solvers, such as the MATLAB built-in function lyapchol, directly computing the searched for Cholesky factors T_α. The additional computations forcing the stability of the reduced SLS are also performed in advance of the simulation and therefore are part of the offline phase. The entire procedure is sketched in Algorithm 4.1.

Note that the method described above relies on the reduction of the k subsystems of the SLS and therefore becomes numerically expensive for large numbers k. Thus, the entire procedure, of course, is built on a proper approximation of the original

Algorithm 4.1 Stability-preserving MOR for SLS

INPUT: Subsystems $(E_\alpha,\ A_\alpha,\ B_\alpha,\ C_\alpha)$, $\alpha = \{1,\dots,k\}$.
OUTPUT: Truncation matrices $V_\alpha,\ W_\alpha$.

1: **for** $\alpha = 1,\dots,k$ **do**
2: Compute stable $(\hat{E}_\alpha,\ \hat{A}_\alpha,\ \hat{B}_\alpha,\ \hat{C}_\alpha)$ and $V_\alpha,\ W_\alpha$, using stability-preserving, projection-based MOR.
3: Choose $Q > 0$ and solve the small-scale ALE

$$\hat{E}_\alpha^T \hat{P}_\alpha \hat{A}_\alpha + \hat{A}_\alpha^T \hat{P}_\alpha \hat{E}_\alpha = -Q,$$

for $\hat{P}_\alpha > 0$.
4: Compute the factorization $\hat{P}_\alpha = T_\alpha^T T_\alpha$.
5: Compute final truncation matrices $V_\alpha = V_\alpha T_\alpha^{-1}$ and $W_\alpha = W_\alpha T_\alpha^T$.
6: **end for**

full-order LTV system by a SLS, consisting of a reasonable number of full-order LTI subsystems. Further, recall that the reduced-order subsystems may evolve in different subspaces. Hence, a state transformation

$$\tilde{x}_{\alpha^+}(t_s^+) = W_{\alpha^+}^T E_{\alpha^{+/-}} V_{\alpha^-} \tilde{x}_{\alpha^-}(t_s^-) \tag{4.13}$$

performing the switch between the individual subspaces has to be applied during the online simulation. The projection (4.13) describes the realignment to the new basis of the reduced state vector $\tilde{x}_\alpha$ with respect to the switching from subspaces $V_{\alpha^-}, W_{\alpha^-}$ to $V_{\alpha^+}, W_{\alpha^+}$ at switching time t_s. Note that for $E_{\alpha^{+/-}}$ either E_{α^+} or E_{α^-} is used for the state transformation. This depends on the definition of

$$\Pi_\alpha := \begin{cases} V_\alpha \left((W_\alpha^T E_\alpha) V_\alpha\right)^{-1} (W_\alpha^T E_\alpha) \\ (E_\alpha V_\alpha) \left(W_\alpha^T (E_\alpha V_\alpha)\right)^{-1} W_\alpha^T \end{cases},$$

being the generalized projector between the corresponding subspaces. In order to avoid the computation of the reduced state transformation $W_{\alpha^+}^T E_{\alpha^{+/-}} V_{\alpha^-}$ at every switching instance, the products should be pre-computed. This leads to an increasing offline computation time and storage amount but accelerates the online simulation time significantly.

Note that the reduction of the single subsystems may result in different reduced orders r_α whereas, the presented procedure, in particular ensuring the stability of the reduced-order SLS, is based on reduced-order models and the associated matrices P_α being of the same reduced orders $r_\alpha = r$. That is, the reduced orders of the different submodels, in a final step, need to be adjusted. This in direct consequence, may lead to unnecessarily large submodels $(\hat{E}_\alpha,\ \hat{A}_\alpha,\ \hat{B}_\alpha,\ \hat{C}_\alpha)$.

Algorithm 4.2 Iterative rational Krylov algorithm (IRKA)

INPUT: System (E, A, B, C), initial sets of interpolation points s_i and tangential directions b_i, c_i, $i = 1, \ldots, r$ that are each closed under conjugation and a convergence tolerance *tol*.

OUTPUT: Reduced-order system $(\hat{E}, \hat{A}, \hat{B}, \hat{C})$ and / or V, W.

1: Compute V and W such that

$$V = \text{span}\left\{(s_1 E - A)^{-1} B b_1, \ldots, (s_r E - A)^{-1} B b_r\right\},$$
$$W = \text{span}\left\{(s_1 E - A)^{-T} C^T c_1, \ldots, (s_r E - A)^{-T} C^T c_r\right\}.$$

2: **while** relative change in $s_i >$ *tol* **do**

3: $\quad \hat{E} = W^T E V, \hat{A} = W^T A V, \hat{B} = W^T B, \hat{C} = C V.$

4: $\quad$ Compute (λ_i, x_i, y_i^*) with $y_j^* \hat{E} x_i = \delta_{ij}$, where $\lambda_i \in \Lambda(\hat{A}, \hat{E})$ and x_i, y_i are the left and right eigenvectors, respectively, associated to λ_i.

5: $\quad$ Assign $s_i \leftarrow -\lambda_i$, $b_i \leftarrow \hat{B}^T y_i$ and $c_i \leftarrow \hat{C} x_i$, for $i = 1, \ldots, r$.

6: $\quad$ Update V and W such that

$$V = \text{span}\left\{(s_1 E - A)^{-1} B b_1, \ldots, (s_r E - A)^{-1} B b_r\right\},$$
$$W = \text{span}\left\{(s_1 E - A)^{-T} C^T c_1, \ldots, (s_r E - A)^{-T} C^T c_r\right\}.$$

7: **end while**

8: $\hat{E} = W^T E V, \hat{A} = W^T A V, \hat{B} = W^T B, \hat{C} = C V.$

4.1.2. MOR for Parametric LTI Systems

In this section, we assume the time-dependence of system (4.1) to be a continuous parameter μ, that implicitly depends on time. Depending on μ, the actual LTV system (4.1) becomes the generalized LPV state-space system

$$\begin{aligned} E(\mu)\dot{x} &= A(\mu)x + B(\mu)u, \\ y &= C(\mu)x \end{aligned} \tag{4.14}$$

with $E(t) \leftarrow E(\mu)$, $A(t) \leftarrow A(\mu)$, $B(t) \leftarrow B(\mu)$, and $C(t) \leftarrow C(\mu)$. As stated in Section 2.2, the main task of projection based model order reduction is to find appropriate truncation matrices V, W. In case of PMOR, the desired goal is to determine a set of global truncation matrices valid over the entire parameter range, such that the physical interpretability of the full-order parametric model (4.14) is preserved.

Although, parametric model order reduction is a quite young field of research, there is already a number of developments based on multivariate Padé approximation, see e.g., [205, 71, 73, 138, 63, 69, 72]. These schemes are also called moment matching methods and the several contributions mainly differ in the way moments are determined in order to achieve satisfactory approximations. Disadvantageously, procedures using explicitly computed moments suffer from the same numerical instabilities as their non-

Algorithm 4.3 Piecewise $\mathcal{H}_2$ optimal interpolatory PMOR

INPUT: Parameter sample points $\{\mu_1, \ldots, \mu_k\}$ and reduction orders $\{r_1, \ldots, r_k\}$.
OUTPUT: Global truncation matrices V and W.

1: **for** $i = 1, \ldots, k$ **do**
2: Apply Algorithm 4.2 to $(E(\mu_j), A(\mu_j), B(\mu_j), C(\mu_j))$ to construct $V_j,\ W_j \in \mathbb{R}^{n \times r_j}$.
3: **end for**
4: Concatenate V_j and W_j for $j = 1, \ldots, k$ to obtain the global truncation matrices $V, W \in \mathbb{R}^{n \times r}$ with $r = \sum_{j=1}^{k} r_j$:

$$V = [V_1, \ldots, V_k] \qquad W = [W_1, \ldots, W_k].$$

5: Use an SVD or rank-revealing QR factorization to remove rank-deficient components from V and W.

parametric counterparts. On the other hand, implicit approaches can overcome these problems but may be unsuitable for large numbers of parameters. Other approaches, dealing with the direct interpolation of the full transfer function, are considered in, e.g., [15]. A number of contributions regarding reduced-basis methods for parametric models can be found in, e.g., [12, 87, 93] and PMOR techniques using matrix interpolation are presented in, e.g., [4, 155]. A recent survey on parametric model order reduction approaches can be found in [21].

Here, we follow the procedure in [14, Section 5]. The main idea is to choose a set of parameter sample points $\mu_1, \ldots, \mu_k$ and to determine a pair of truncation matrices $V_j,\ W_j$ for a fixed parameter point μ_j. Defining the transfer function $\mathcal{H}^{(j)}(s) := \mathcal{H}(s, \mu_j)$, for each $j = 1, \ldots, k$, the system associated to $\mathcal{H}^{(j)}(s)$ is in fact a non-parametric full-order LTI model. Hence, in general any MOR method suitable for approximating the full-order model can be employed. Here, we use the iterative rational Krylov algorithm [90] that, as mentioned before, provides an automatic shift selection strategy. Moreover, each of the generated reduced-order models defines an locally $\mathcal{H}_2$-optimal approximation of the respective full-order system for $\mu_j,\ 1, \ldots, k$. A sketch of IRKA for generalized MIMO systems is given in Algorithm 4.2 [14, Algorithm 3.1].

An optimal $\mathcal{H}_2$ approximation $\hat{\mathcal{H}}$ is the solution to the minimization problem

$$\min_{\hat{\mathcal{H}}} \|\mathcal{H} - \hat{\mathcal{H}}\|_{\mathcal{H}_2} \quad \text{with} \quad \|\mathcal{H}\|_{\mathcal{H}_2} := \left(\frac{1}{2\pi} \int_{-\infty}^{\infty} \|\mathcal{H}(\jmath\omega)\|_F^2 \, d\omega \right)$$

and $\|.\|_F$ being the Frobenius norm of a matrix. Given the locally $\mathcal{H}_2$-optimal reduced-order models of dimension r_j, at the selected parameter sample points μ_j, defined by the pairs $V_j,\ W_j \in \mathbb{R}^{n \times r_j}$, $j = 1, \ldots, k$, a pair of global projection matrices

$$V = \left[V_1, \ldots, V_k\right], \quad W = \left[W_1, \ldots, W_k\right] \in \mathbb{R}^{n \times r} \tag{4.15}$$

is obtained by concatenation with $r = \sum_{j=1}^{k} r_j$. For better numerical properties it is recommended to additionally orthogonalize the truncation matrices V, W. The entire procedure is given in Algorithm 4.3 [14, Algorithm 5.1]. Note that the resulting global matrices V, W in general do not generate a globally $\mathcal{H}_2$-optimal reduced-order model although each of the subsystems is locally $\mathcal{H}_2$-optimal. Still, the transfer function $\hat{\mathcal{H}}$, associated to the globally reduced-order model, interpolates $\mathcal{H}$ at all parameter choices. Further note that the exact interpolation property will be lost if a truncated SVD is to be used in Step 5 of Algorithm 4.3. Details are given in [14].

Given the global projection bases V, W the reduced-order matrices

$$\begin{aligned} \hat{E}(\mu) &= W^T E(\mu) V, & \hat{A}(\mu) &= W^T A(\mu) V, \\ \hat{B}(\mu) &= W^T B(\mu) & \hat{C}(\mu) &= C(\mu) V, \end{aligned}$$

can be computed. This requires the computation of $\hat{E}(\mu)$, $\hat{A}(\mu)$, $\hat{B}(\mu)$, $\hat{C}(\mu)$ for each parameter value μ within the online simulation of the system and therefore becomes a highly time consuming procedure. Hence, an efficient splitting of the computations into an offline and online part is considered. That is, the parameter-dependence of the system matrices $\hat{E}(\mu)$, $\hat{A}(\mu)$, $\hat{B}(\mu)$, $\hat{C}(\mu)$ is rewritten in an affine form

$$\begin{aligned} E(\mu) &= E_0 + e_1(\mu)E_1 + \cdots + e_{m_E}(\mu)E_{m_E}, \\ A(\mu) &= A_0 + f_1(\mu)A_1 + \cdots + f_{m_A}(\mu)A_{m_A}, \\ B(\mu) &= B_0 + g_1(\mu)B_1 + \cdots + g_{m_B}(\mu)B_{m_B}, \\ C(\mu) &= C_0 + h_1(\mu)C_1 + \cdots + h_{m_C}(\mu)C_{m_C}, \end{aligned} \tag{4.16}$$

where $m_E,\ m_A,\ m_B,\ m_C \in \mathbb{N}$. Note that the numbers of summands m_E, m_A, m_B, and m_C do not necessarily have to be equal. Such a parameter affine representation can always be achieved, see e.g., [92]. Given this form, the parameter dependent matrices $\hat{E}(\mu)$, $\hat{A}(\mu)$, $\hat{B}(\mu)$, and $\hat{C}(\mu)$ can be reduced in the form

$$\begin{aligned} \hat{E}(\mu) &= \sum_i^{m_E} e_i(\mu) W^T E_i V, & \hat{A}(\mu) &= \sum_i^{m_A} f_i(\mu) W^T A_i V, \\ \hat{B}(\mu) &= \sum_i^{m_B} g_i(\mu) W^T B_i, & \hat{C}(\mu) &= \sum_i^{m_C} h_i(\mu) C_i V, \end{aligned} \tag{4.17}$$

where the computation of $W^T E_i V$, $W^T A_i V$, $W^T B_i$, $C_i V$ is now parameter independent and therefore performed in the offline stage. Using the affine representation (4.16), the structure of the parameter-dependence is automatically preserved and the online evaluation of the parameter is reduced to the evaluation of the functions e_i, f_i, g_i, h_i, $i = 1, \ldots, m_E/m_A/m_B/m_C$. These functions may be linear or nonlinear, but are assumed to be smooth enough to allow for interpolation. Moreover, the representation (4.16) is not unique. In general, there may be many ways to define such an affine splitting. That is, in particular the parameter representing functions e_i, f_i, g_i, h_i and the number of summands m_E, m_A, m_B, m_C may vary with respect to the chosen representation. In order to also keep the computational effort of the offline part in appropriate limits, small numbers m_E, m_A, m_B, $m_C \ll n$ are desired.

4.2. Balanced Truncation for Linear Time-Varying Systems

In Section 2.2.1 the general idea of balanced truncation model order reduction for linear time-invariant systems was stated. This section reviews its extension to the linear time-varying case, originally developed in [183, 202, 171]. Therein it is shown, that the preservation of stability and the existence of an error bound, well known for LTI systems, can be extended to the LTV case. In contrast to the LTI case, balanced truncation for LTV systems relies on the time-dependent reachability and observability Gramians $P(t)$ and $Q(t)$ defined as the solutions of the associated generalized differential Lyapunov equations. Note that both, the SLS and PMOR approaches for the approximation of the LTV structure, are presented for the general case, where $E = E(t)$. For the same reason as highlighted in Equation (2.11), here we restrict to non-varying matrices E such that the generalized reachability and observability DLEs read

$$E\dot{P}(t)E^T = A(t)P(t)E^T + EP(t)A(t)^T + B(t)B(t)^T, \quad P(t_0) = 0, \tag{4.18}$$

$$-E^T\dot{Q}(t)E = A(t)^TQ(t)E + E^TQ(t)A(t) + C(t)^TC(t), \quad Q(t_f) = 0. \tag{4.19}$$

The zero initial and final conditions of the respective DLEs are given from the alternate integral representations

$$P(t) = \int_{t_0}^{t} \Phi_{E^{-1}A}(t,\tau)E^{-1}B(\tau)B(\tau)^TE^{-T}\Phi_{E^{-1}A}(t,\tau)^T d\tau,$$

$$Q(t) = \int_{t}^{t_f} \Phi_{E^{-1}A}(\tau,t)^TC(\tau)^TC(\tau)\Phi_{E^{-1}A}(\tau,t) d\tau$$

with the transition matrix $\Phi_{E^{-1}A}(t,\tau)$ of the underlying LTV system (4.1). By definition, the Gramians $P(t)$ and $Q(t)$ are symmetric and for $B(t)B(t)^T$ and $C(t)^TC(t)$ being positive/negative semi-definite, the same holds for $P(t)$ and $Q(t)$. Provided that the LTV system is in addition reachable, observable, and the time-dependent matrices $A(t)$, $B(t)$, and $C(t)$ are continuous and bounded, the Gramians $P(t)$ and $Q(t)$ are positive definite for all $t \in [t_0, t_f]$. Therefore, the product $P(t)E^TQ(t)E$ has real positive eigenvalues $\lambda_i\big(P(t)E^TQ(t)E\big)$ and according to the LTI case, the time-dependent Hankel singular values associated to system (4.1) are defined as

$$\tilde{\sigma}_i(t) = \sqrt{\lambda_i\big(P(t)E^TQ(t)E\big)}, \qquad i = 1,\dots,n,\ t \in [t_0, t_f]. \tag{4.20}$$

In analogy to Section 2.2.1, an LTV system (4.1) is called balanced if the time-dependent Gramians satisfy $P(t) = E^TQ(t)E = \Sigma(t) = \operatorname{diag}(\sigma_1(t),\dots,\sigma_n(t))$ for all $t \in [t_0, t_f]$.

Algorithm 4.4 Balanced truncation for LTV systems.

Input: $\big(E, A(t), B(t), C(t)\big)$, truncation tolerance BT_{tol}

Output: a reduced-order system $\big(\hat{E}, \hat{A}(t), \hat{B}(t), \hat{C}(t)\big)$

1: Compute the Cholesky factors $R(t)$ and $L(t)$ of the reachability and observability Gramians $P(t) = R(t)R(t)^T$ and $Q(t) = S(t)S(t)^T$ satisfying the DLEs (4.18) and (4.19), respectively.

2: Compute the singular value decomposition

$$S(t)^T ER(t) = [\, U_1(t),\ U_2(t)\,] \begin{bmatrix} \Sigma_1(t) & \\ & \Sigma_2(t) \end{bmatrix} [\, Z_1(t),\ Z_2(t)\,]^T,$$

where the matrices $[\, U_1(t),\ U_2(t)\,]$ and $[\, Z_1(t),\ Z_2(t)\,]$ have orthonormal columns, $\Sigma_1(t) = \mathrm{diag}(\sigma_1(t), \dots, \sigma_r(t))$ and $\Sigma_2(t) = \mathrm{diag}(\sigma_{r+1}(t), \dots, \sigma_n(t))$ with $\sigma_{r+1} > BT_{tol}$.

3: Compute the reduced-order system (4.2) with

$$\hat{E}(t) = W(t)^T EV(t), \quad \hat{A}(t) = W(t)^T\big(AV(t) - E\dot{V}(t)\big),$$
$$\hat{B}(t) = W(t)^T B(t), \quad \hat{C}(t) = C(t)V(t),$$

where $V(t) = R(t)Z_1(t)\Sigma_1(t)^{-1/2}$ and $W(t) = S(t)U_1(t)\Sigma_1(t)^{-1/2}$.

In [183, Section III], for standard systems $(t; A, B, C)$ it is shown that a balancing transformation with unsorted $\tilde{\Sigma}(t)$, based on an eigendecomposition of the product of the Gramians, exists and further is unique if the continuous differentiable eigenvalues $\tilde{\sigma}_i(t)$, $i = 1, \dots, n$, of the Gramian product are distinct except for a finite number of intersection points. Given the balanced realization, the dominant subsystem, constituting the sought for reduced-order system can be obtained by simply reordering the states, with respect to the magnitude of the HSVs $\tilde{\sigma}_i(t)$, by a permutation. Since the trajectories of two HSVs $\tilde{\sigma}_i(t)$ and $\tilde{\sigma}_j(t)$ for $i \neq j$ may intersect and thus swap their roles within the ordered sequence $\Sigma(t)$, the permutation in the above procedure needs to be constant. This limits the method to a predefined minimal reduced order, covering the dominant states for the entire simulation interval. In fact, the determination of such a minimal necessary order is a challenging task and may lead to reduced-order models of unnecessarily "large" dimensions at certain time instances $t \in [t_0, t_f]$.

However, assume that the system (4.1) is stable, reachable, and observable and therefore the Gramians are symmetric positive definite, for all t. Then, following the square-root balanced truncation approach for LTI systems, for the Gramians $P(t)$, $Q(t)$, there exist (Cholesky(-like)) factorizations

$$P(t) = R(t)R(t)^T, \qquad Q(t) = S(t)S(t)^T, \ \forall t$$

and hence the implicitly balancing truncation matrices read

$$V(t) = R(t)Z_1(t)\Sigma_1(t)^{-\frac{1}{2}}, \qquad W(t) = S(t)U_1(t)\Sigma_1(t)^{-\frac{1}{2}}, \tag{4.21}$$

where, similar to the time-invariant case, the matrices $\Sigma_1(t)$, $U_1(t)$, and $Z_1(t)$ originate from the SVD

$$S(t)^T ER(t) = U(t)\Sigma(t)Z(t)^T = \begin{bmatrix} U_1(t) & U_2(t) \end{bmatrix} \begin{bmatrix} \Sigma_1(t) & \\ & \Sigma_2(t) \end{bmatrix} \begin{bmatrix} Z_1(t)^T \\ Z_2(t)^T \end{bmatrix}. \tag{4.22}$$

A sketch of the balanced truncation square-root procedure is given in Algorithm 4.4. As for the time-invariant BT procedure, the main step, in computing suitable truncation matrices, is to solve the reachability and observability Lyapunov equations, given in terms of the DLEs (4.18), and (4.19) associated to the time-varying system (4.1). In Chapter 5 a number of time integration schemes will be discussed. Details on the numerical implementation, in particular for large-scale problems, are given in Chapter 6.

Note that even if the eigenvalues λ_i, $i = 1, \ldots, n$, of the product PE^TQE and consequently every individual singular value $\tilde{\sigma}_i(t)$ from (4.20) is a continuously differentiable function, this may not hold for the sequence of the ordered Hankel singular values $\Sigma(t)$, obtained by the SVD (4.22). That is, in general, $V(t) = R(t)Z_1(t)\Sigma_1(t)^{-\frac{1}{2}}$ is not continuously differentiable. Moreover, the reduced-order states, associated to the truncation matrices $V(t)$, $W(t)$ from Equation (4.21), may live in different subspaces and therefore, as for the SLS from Section 4.1.1, within the forward simulation a state transformation needs to be applied at every time step, taking care of the swapping HSVs and the associated change of subspaces. Therefore, for the approximate representation $x(t) \approx V(t)\hat{x}(t)$ of the original state, we consider the discretization

$$\frac{d}{dt}x(t_k) = \frac{d}{dt}\Big(V(t_k)\hat{x}(t_k)\Big) \approx \frac{1}{\tau_k\beta}\sum_{j=0}^{s} \alpha_j V_{k-j}\hat{x}_{k-j},$$

of its derivative, as it is known from the BDF methods. This time discretization yields the linear reduced-order system

$$(\alpha_0\hat{E}_k - \tau_k\beta\hat{A}_k)\hat{x}_k = -W_k^T E\sum_{j=1}^{s} \alpha_j V_{k-j}\hat{x}_{k-j} + \tau_k\beta\hat{B}_k u_k, \tag{4.23}$$

where the expression $W_k^T E \sum_{j=1}^{s} \alpha_j V_{k-j}\hat{x}_{k-j}$ realizes the state transformation by implicitly lifting the previous time solutions $\hat{x}_{k-j}$ with V_{k-j} to the full-order state-space and then again projecting onto the subsequent low-dimensional subspace by $W_k^T E$. This automatically allows us to vary the system dimension r of the reduced-order model at every time step.

Still, storing the truncation matrices $V(t)$, $W(t)$ at a set of discrete time instances $\mathcal{T} = \{t_0, t_1, \ldots, t_f\}$ with $t_0 < t_1 < \cdots < t_f$, turns out to be highly memory consuming. This, in particular, is a serious problem for stiff DLEs that demand for small time step sizes and consequently yield a significant number of discrete truncation matrices $V(t_k)$, $W(t_k)$, $t_k \in \mathcal{T}$. Moreover, given the time-dependence in terms of the system matrices E, $A(t)$, $B(t)$, $C(t)$, as well as the truncation matrices $V(t)$, $W(t)$ requires the reduced-order matrices (4.3) to be computed at every step of the online forward simulation of the dynamical system. Considering the time-variability to be a special case of a parameter-dependence, the reduced-order model can be computed in the offline phase, using, e.g., an affine-representation of the system matrices together with a computation of constant, global truncation matrices V, W valid on the entire time interval, similar to the parameter-affine representation (4.16) and the concatenated projection matrices (4.15), respectively. In that sense, the BT procedure for LTV systems is similar to the parametric reduction framework presented in Section 4.1.2, whereas the sample points are chosen by the time discretization used for the solution of the reachability and observability DLEs. However, using global truncation matrices the stability of the ROM and the existence of an error bound can no longer be guaranteed. Still, in the special case that A is symmetric negative definite and E is positive definite, the stability can be ensured, using a one-sided projection, i.e., $V = W$. This also holds for the PMOR scheme. Further note that, given V constant, we have $\dot{V} = 0$. That is, no state-transformation, due to varying subspaces is needed within the online phase.

Another, critical issue from the numerical point of view are the zero initial and final condition of the reachability and observability DLEs, respectively. That is, the Gramians $P(t_0) = Q(t_f) = 0$ lead to truncation matrices $V(t_0) = V(t_f) = W(t_0) = W(t_f) = 0$ and therefore do not allow the computation of a non-zero ROM at t_0 and t_f. Now, consider the forward simulation of a ROM (4.2) and the first step of the integration scheme (4.23) with its right hand side $-W_1^T E\alpha_1 V_0 \hat{x}_0 + \tau_1 \beta \hat{B}_1 u_1$ and $V_0 = V(t_0) = 0$. Here, the state-transformation $W_1^T E\alpha_1 V_0 \hat{x}_0$ is of particular interest. Given non-zero initial conditions x_0, and $\hat{x}_0$, the transformation will lift $\hat{x}_0$ to $0 = \tilde{x}_0 = V_0 \hat{x}_0 \neq x_0$ and thus, the ROM starts at a zero initial condition, even if the actual starting point is non-zero. For the special case, where a zero initial condition is given, this obviously, is not a problem. For conviction, see Section 4.3.3. Still, in any case, the problem is present for $t = t_f$, such that the final ROM, being zero, is meaningless.

4.3. Numerical Experiments

4.3.1. Moving Load Problem: Machine Stand-Slide Structure

To illustrate the application of the model reduction methods shown above, as a first example, we consider the machine stand example given in [89]. Therein, the system variability is induced by a moving tool slide on the guide rails of the stand (see Figure 4.2). The aim is to determine the thermally driven displacement of the machine stand structure. Following the model setting in [89], the deformation is assumed to not have any effect on the thermal behavior. This leads to a one-sided coupling of the deformation and heat models. Since the model equations for the slide structure result in a "simple" LTI system, here we restrict our considerations to the stand model.

Figure 4.2.: Tool slide (left) and machine stand (right) geometries.

Therefore, we consider the heat equation

$$\begin{aligned}
c_p\rho\dot{\Theta} &= \operatorname{div}(\lambda\nabla\Theta), && \text{on } \Omega && (4.24)\\
\lambda\frac{\partial}{\partial n}\Theta &= \kappa^{ext}(\Theta^{ext} - \Theta), && \text{on } \Gamma^{ext} \subset \partial\Omega && (4.25)\\
\lambda\frac{\partial}{\partial n}\Theta &= \kappa^{temp}(\Theta^{temp} - \Theta), && \text{on } \Gamma^{temp} \subset \partial\Omega && (4.26)\\
\lambda\frac{\partial}{\partial n}\Theta &= q_{fric} + \kappa^{c}(\Theta^{c} - \Theta), && \text{on } \Gamma^{c} \subset \partial\Omega && (4.27)\\
\Theta(0) &= \Theta_0, && \text{at } t = 0 && (4.28)
\end{aligned}$$

where Θ^c is the temperature of the contact area of the tool slide, Θ^{ext} denotes the exterior temperature, c_p is the material specific heat capacity, ρ the density and λ describes the heat conductivity of the stand. The boundary Γ^c denotes the time-varying contact boundary, which will vary with the position of the slide on the stand surface. The boundary condition, stated in Equation (4.25), describes the heat exchange of the stand structure with the ambient. Equation (4.26) represents a temperature control with constantly applied temperatures at the bottom and the right wall of the stand structure. Equation (4.27) models the temperature exchange of the stand and the moving slide at the contact boundary Γ^c. In addition to the heat amount exchanged, the latter includes the friction induced thermal load

$$q_{fric} = \frac{1}{2}\frac{\eta v(t)^2}{\delta s},$$

caused by the slide movement. Here, v is the velocity (feed rate) of the moving slide, η is the dynamical viscosity, and δs denotes the width of the gap between the two structures.

Using a finite element (FE) discretization and denoting the external influences, i.e., the friction driven portion, the heat transferred from the slide at the contact boundary, and the external temperatures, as the system input z, we obtain the dynamical heat model

$$E_{th}\dot{T}(t) = A_{th}(t)T(t) + B_{th}(t)z(t). \tag{4.29}$$

Here, $T(t) \in \mathbb{R}^{n_{th}}$ denotes the discrete, deformation independent temperature field, $E_{th} \in \mathbb{R}^{n_{th}\times n_{th}}$ is the FE mass matrix, and the varying (with respect to the boundary Γ^c) matrices $A_{th}(t) \in \mathbb{R}^{n_{th}\times n_{th}}$, $B_{th}(t) \in \mathbb{R}^{n_{th}\times m}$ are the system and input matrices, respectively. The particular form of the input $z \in \mathbb{R}^m$ depends on the MOR ansatz we choose and is therefore described in the corresponding sections below. As one can easily check, in this example the variability only affects the system and input matrices $A_{th}(t)$ and $B_{th}(t)$. In order to determine the elastic displacement field of the stand, we additionally need an elasticity model. Since the mechanical behavior of the machine stand is much faster than the propagation of the thermal field, it is sufficient to define a stationary linear elasticity model. That is, considering the balance of forces

$$-\operatorname{div}\sigma(\varepsilon(u), \Theta) = f \quad \text{in } \Omega$$

and together with an additive split of the stress tensor σ into its mechanically and thermally induced parts and the usual homogeneous and isotropic stress-strain relation, we obtain the constitutive law

$$\begin{aligned}
-\operatorname{div}\sigma(\varepsilon(u), \Theta) &= f, \\
\sigma(\varepsilon(u), \Theta) &= \sigma^{\mathrm{el}}(\varepsilon(u)) + \sigma^{\mathrm{th}}(\Theta), \\
\sigma^{\mathrm{el}}(\varepsilon(u)) &= \frac{E}{1+\nu}\varepsilon(u) + \frac{E\nu}{(1+\nu)(1-2\nu)}\operatorname{tr}(\varepsilon(u))\, I_d, \\
\sigma^{\mathrm{th}}(\Theta) &= -\frac{E}{1-2\nu}\beta\,(\Theta - \Theta_{ref})I_d
\end{aligned} \tag{4.30}$$

to describe the displacement field u. For a detailed derivation and further information on the continuum mechanics related issues, yielding (4.30), we refer to [37, Section 1.12] and [67, Section 2.8]. However, $\varepsilon(u) = \frac{1}{2}(\nabla u + \nabla u^T)$ denotes the linearized strain-tensor on the domain Ω. The expression f describes external body forces inducing elastic deformations. In the following example these forces are assumed to be zero. The coefficients E_u, ν, β describe Young's modulus, Poisson's ratio and the thermal expansion coefficient, respectively, depending on the material. The variable I_d is the identity of dimension d. Here $d = 3$ is the spatial dimension of Ω. Moreover, the geometrical dimensions and the specific thermal and elastic material data of the stand and slide structures are given in Table 4.1. The coupling of the thermal and elastic model is given by the expression

$$\beta(\Theta - \Theta_{ref})I_d$$

which describes the thermally driven distortion of the domain Ω. That is, the resulting displacement is induced by the change of temperature Θ with respect to a given

Parameter	Unit	Stand	Slide
$(H \times W \times D)$	mm	$(2010 \times 519 \times 480)$	$(500 \times 430 \times 490)$
c_p	$\frac{Ws}{kgK}$	460	
ρ	$\frac{kg}{m^3}$	7200	
λ	$\frac{W}{mK}$	50	
κ^c	$\frac{W}{m^2K}$	50	
κ^{ext}	$\frac{W}{m^2K}$	10	
κ^{temp}	$\frac{W}{m^2K}$	100	
β	$\frac{1}{K}$	1.1e-5	
ν	−	0.28	
E_u	$\frac{N}{m^2}$	169e9	
η	$\frac{Ns}{m^2}$	0.1980	
δs	mm	1e-2	

Table 4.1.: Geometrical dimensions and material parameters of the stand and slide structures.

reference temperature Θ_{ref} of Ω at time t_0. Using the same spatial FE discretization as for the heat model (4.29), we obtain the algebraic linear system

$$0 = A_{thel}T(t) - A_{el}u(t), \tag{4.31}$$

where $A_{thel} \in \mathbb{R}^{n_{el} \times n_{th}}$ denotes the coupling of the discrete temperature field T in direction of the displacement field $u \in \mathbb{R}^{n_{el}}$ and $A_{el} \in \mathbb{R}^{n_{el} \times n_{el}}$ is the elasticity system matrix. Since, the influence of the deformation u in direction of the thermal behavior is neglected in the model (4.24)-(4.25), there is no coupling matrix $A_{elth} \in \mathbb{R}^{n_{th} \times n_{el}}$. The dimension $n_{el} = 3n_{th}$ denotes the number of elastic degrees of freedom in the FE nodes with respect to the three spatial directions. Assuming the deformation to be small, we consider only translational degrees of freedom. Note that the deformation u is not affected by the movement of the tool slide explicitly and therefore the matrices A_{thel}, A_{el} are constant with respect to the structural variability. Finally, combining Equations (4.29) and (4.31), we end up with the coupled thermo-elastic system

$$\begin{bmatrix} E_{th} & 0 \\ 0 & 0 \end{bmatrix} \begin{bmatrix} \dot{T}(t) \\ \dot{u}(t) \end{bmatrix} = \begin{bmatrix} A_{th}(t) & 0 \\ A_{thel} & -A_{el} \end{bmatrix} \begin{bmatrix} T(t) \\ u(t) \end{bmatrix} + \begin{bmatrix} B_{th}(t) \\ 0 \end{bmatrix} z(t), \tag{4.32}$$

$$\Leftrightarrow \quad E\dot{x}(t) = A(t)x(t) + B(t)z(t).$$

Due to the non-singularity of A_{el}, System (4.32) is an index-1 DAE. Since we are interested in the thermally driven deformations at certain points, we consider the

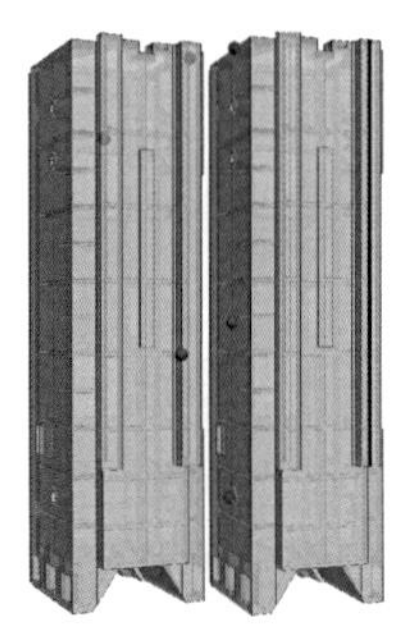

Figure 4.3.: Output locations on the stand structure.

output equation

$$y(t) = \begin{bmatrix} 0, & C_{el} \end{bmatrix} \begin{bmatrix} T(t) \\ u(t) \end{bmatrix} = Cx(t) \tag{4.33}$$

with $C \in \mathbb{R}^{q \times n}, n = n_{th} + n_{el} = 4n_{th}$ and $C_{el} \in \mathbb{R}^{q \times n_{el}}$, which only filters the deformation information, we are interested in. In practice, such points might be the tool center point or connections to neighboring assembly groups. The numerical results, presented in Section 4.3.1, are given at the locations depicted in Figure 4.3. The outputs are numbered by 1 to 3 from top to bottom for the left stand picture and 4 to 6 from top to bottom for the right stand graphic.

Using an appropriately refined FE discretization, the dimension n of the system becomes very large and thus, we aim at full utilization of the special differential-algebraic structure of the coupled thermo-elastic system (4.32). The application of the Schur complement to Equations (4.32) and (4.33) exploits the one-sided coupling (see e.g., [77]). That is exploiting the zero block in the upper right corner of $A(t)$, we obtain

$$\begin{aligned} E_{th}\dot{T} &= A_{th}(t)T + B_{th}(t)z, \\ y &= C_{el}A_{el}^{-1}A_{thel}T = \bar{C}T. \end{aligned} \tag{4.34}$$

The MOR techniques now have to be applied to system (4.34) of dimension n_{th} which is of the same structure as system (4.32), (4.33) of dimension n. That is, using the same FE discretization for the thermal and elastic model, we a priori reduce the system dimension from $n = 4n_{th}$ to n_{th}. Again, the particular form of $A(t)$, $B(t)$ or $A_{th}(t)$, $B_{th}(t)$, respectively, will slightly vary with respect to the system interpretation and therefore will be described in the corresponding sections below. We consider a time horizon of 16.5 hours, representing a realistic operation time for machining processes. The tool slide trajectory is defined by the velocity profile, presented in Figure 4.4(a), starting at the lowest position possible. The FE grid consists of 16 626 nodes and at the same time defines the number n_{th} of thermal degrees of freedom. Moreover, the surrounding

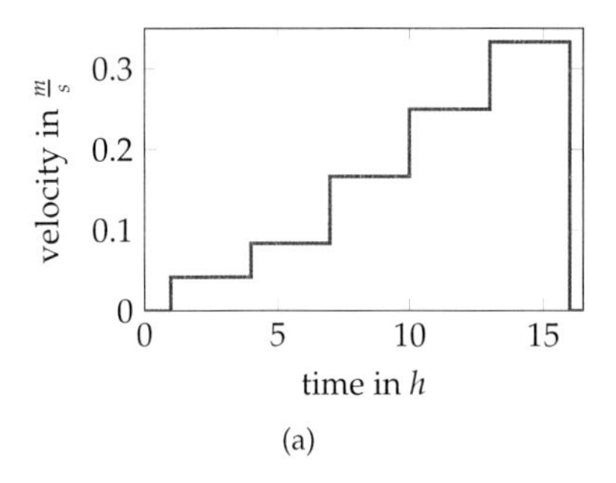

(a)

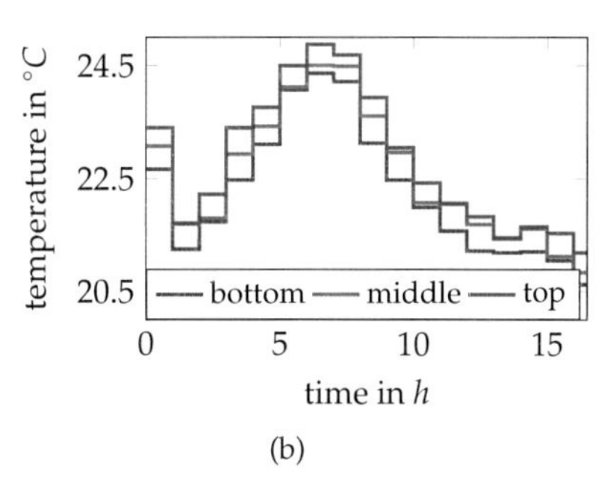

(b)

Figure 4.4.: Velocity profile of the tool slide movement (a) and evolution of temperature for horizontal air thresholds (b).

of the stand is divided into three equally sized horizontal areas, modeling different temperature thresholds. The corresponding trajectories

$$T^{ext} = \begin{cases} T^{ext}_{top}(t), & y > 1340\ mm, \\ T^{ext}_{mid}(t), & 670\ mm < y \leq 1340\ mm, \\ T^{ext}_{bot}(t), & y \leq 670\ mm, \end{cases} \tag{4.35}$$

are presented in Figure 4.4(b). Furthermore, for the tempering boundary we have $\Gamma^{temp} = \Gamma^{temp}_{bot} \cup \Gamma^{temp}_{rw}$, defined by the bottom Γ^{temp}_{bot} and the right wall Γ^{temp}_{rw}, respectively. For the simulation the constant temperature values $T^{temp}_{bot} = 20\ °C$ and $T^{temp}_{rw} = 22\ °C$ are given.

Due to the size of the FE discretization, the large time horizon, the computational expensive BT procedure for LTV systems would lead to an inappropriate computational time and storage consumption. Therefore, here we only consider the PMOR and SLS approach. A comparison of all MOR techniques, described above, is presented in Section 4.3.3. As mentioned earlier, the PMOR- and SLS-MOR strategies require specific formulations of the time-varying matrices $A_{th}(t)$ and $B_{th}(t)$. A detailed description is given below.

Switched Linear System Formulation

Based on the model, given in [89], the guide rails of the machine stand are modeled as 15 equally distributed horizontal segments (see Figure 4.5). Any of these segments is assumed to be completely covered by the tool slide if its midpoint (in y-direction) is covered by the slide. On the other hand, each segment whose midpoint is not covered is treated as not in contact and therefore the slide always covers exactly 5 segments at each time. This in fact allows the stand to reach 11 distinct discrete positions, given

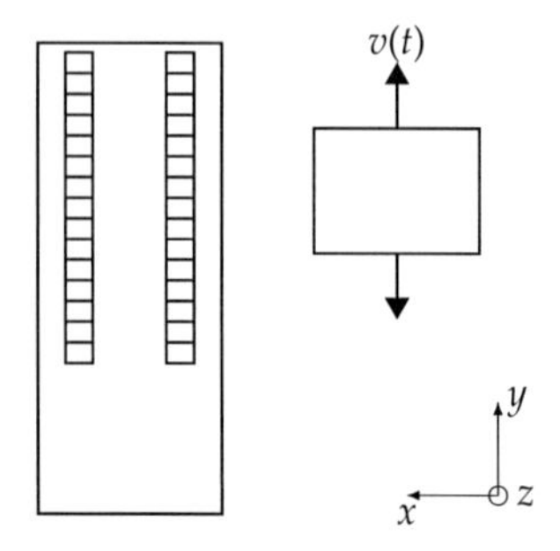

Figure 4.5.: Schematic representation of the partitioned rails and the moving slide.

by the model restrictions. These distinct setups define the subsystems of the switched linear system

$$\begin{aligned} E_{th}\dot{T} &= A_{th}^{\alpha}T + B_{th}z^{\alpha}, \\ y &= \bar{C}T, \end{aligned} \tag{4.36}$$

where α is a piecewise constant function of time, which takes its value from the index set $\mathcal{J} = \{1, \ldots, 11\}$, representing the subsystems (4.36). On the basis of the boundary conditions and the segmentation of the rails, we consider $m = 20$ inputs. For the rail segments, we define

$$z_i^{\alpha} := \begin{cases} z_i, & \text{segment } i \text{ is in contact,} \\ 0, & \text{otherwise,} \end{cases} \qquad i = 1, \ldots, 15. \tag{4.37}$$

where,

$$z_i = q_{fric} + \kappa^c T^c \in \mathbb{R}$$

is the thermal input consisting of the friction induced portion q_{fric} and T^c, being the arithmetic mean of the temperature nodes at the contact surface of the slide. Note that, using the averaged temperature T^c, the single inputs z_i are the same for all segments i in contact. The inputs, representing the tempering boundary condition (4.26), are then $z_{16}^{\alpha} = T_{bot}^{temp}$ and $z_{17}^{\alpha} = T_{rw}^{temp}$. Equation (4.35), representing the ambient thresholds, defines the inputs 18, 19, 20 that are given by $z_{18}^{\alpha} = T_{bot}^{ext}(t)$, $z_{19}^{\alpha} = T_{mid}^{ext}(t)$, and $z_{20}^{\alpha} = T_{top}^{ext}(t)$.

Note that using the input formulation (4.37) allows to remove the variability from the input matrix B_{th} such that the only varying part influencing the model reduction process left in the dynamical system is the system matrix $A_{th}(t) := A_{th}^{\alpha}$. In order to find a locally accurate approximation for each of the subsystems $\alpha \in \mathcal{J}$, we need to compute individual subspaces V_{α}, W_{α}. In order to be able to compute a globally stable reduced order model, we need to preserve the stability of each single system and therefore,

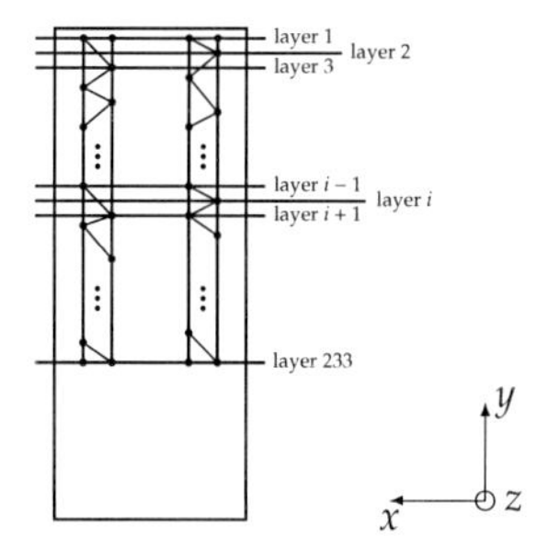

Figure 4.6.: Schematic depiction of the horizontal layers, defining the parameter affine representation, based on FE nodes with the same vertical coordinates.

we use the balanced truncation method. Still, note that the preservation of stability for all subsystems is, in general, not sufficient to ensure stability of the entire SLS (see Section 4.1.1). Still, in this example, the overall SLS resulting from the stable subsystems, obtained by the BT method, is stable without any further computational effort.

Parametric Model Formulation

Recall from Section 4.1.2 that for an efficient implementation of the PMOR approach an parameter affine representation (4.16) of the time-varying matrices $A_{th}(t)$ and $B_{th}(t)$ is required. In contrast to the partitioning of the guide rails into 15 segments, here we consider the variability to be a continuous parameter dependence, and therefore discard the given segmentation exploited in the SLS approach. Here, based on the underlying FE grid, the guide rails, i.e., the boundary Γ^c of the stand, are decomposed into a sequence of horizontal layers, consisting of all FE nodes sharing the same vertical coordinate (see Figure 4.6). That is, the affine representation of the matrices $A_{th}(t) = A_{th}(\mu)$ and $B_{th}(t) = B_{th}(\mu)$ and in particular the number of submatrices A_i, B_i, $i = 1, \dots, m_A/m_B$, associated to the layers, relies on the grid resolution of the contact surface Γ^c. Then, the functions f_i, g_i in Equation (4.16) are equal and of the form

$$f_i(\mu) = g_i(\mu) = \begin{cases} 1, & \text{the } i\text{-th horizontal layer is in contact,} \\ 0, & \text{otherwise.} \end{cases}$$

That is, the functions $f_i(\mu)$, $g_i(\mu)$ serve as indicator functions, activating that parts of Γ^c that are covered by the tool slide, depending on the position μ. The given FE grid and the corresponding discretization lead to a splitting of the guide rails into $m_A = m_B = 233$ disjoint horizontal layers. Given the affine representation, we consider

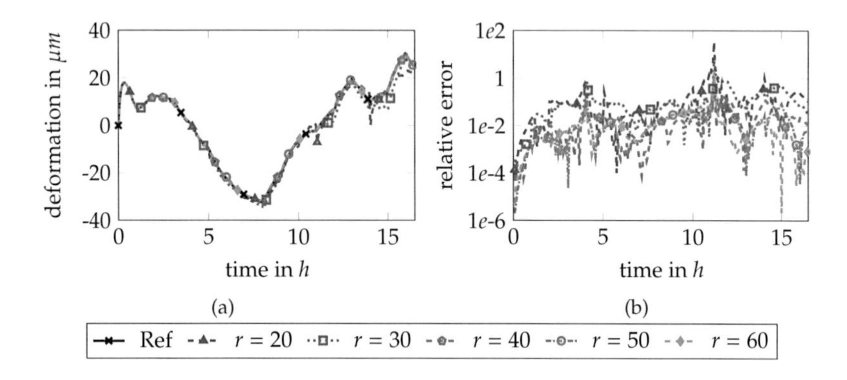

Figure 4.7.: SLS approach: Displacement (a) and corresponding relative errors (b) in output y_1 for reduced orders $r \in 20, 30, 40, 50, 60$.

$m = 6$ inputs, where

$$z_1 = q_{fric} + \kappa^c T^c,$$
$$z_{2,\dots,6} = z^{\alpha}_{16,\dots,20} = [T^{temp}_{bot}, T^{temp}_{rw}, T^{ext}_{bot}, T^{ext}_{mid}, T^{ext}_{top}]^T.$$

That is, the first input represents the friction induced load together with the temperature exchange with the slide, similar to the inputs to the several segments in the SLS approach. Inputs $2, \dots, 6$ are given by the temperature control boundary conditions and the exchange with the ambient, respectively.

Comparison of SLS and PMOR Approach

For the reduction of the subsystems of the SLS, a tolerance of $1e$-9 for the truncation of the HSVs within the balanced truncation MOR method is implemented. The same tolerance is used for the reduction of the slide model. In addition to the truncation tolerance, a maximum reduced order is used, in order to restrict the dimension of the ROMs to a certain limit. The forward simulation of the reduced-order SLS is performed using the implicit Euler scheme with a constant time step size of $\tau = 10$ s. Given the constant step size, the solution operator at every time step of the Euler scheme does not change for the single subsystems. That is, the LU decompositions only have to be computed once for each of the subsystems. In Figure 4.7, we compare the displacement output y_1 for the SLS procedure based on reduced-order submodels of dimension $r = 20, 30, 40, 50, 60$. In Table 4.2, the timings for the SLS based on the submodels of different reduced orders are given. It can be observed that already submodels of dimension $r = 40$ result in a relative error in the percentage range and even below. For this example the reduction and simulation times appear to be

r	Reduction time in s	Simulation time in s
20	880.9862	22.9569
30	898.9794	25.2375
40	903.9425	22.5046
50	896.1119	22.2887
60	897.1985	23.7190

Table 4.2.: SLS approach: Comparison of the reduction and simulation times based on different reduced orders $r \in 20, 30, 40, 50, 60$.

(k, r_j)	$r = \sum r_j$	Reduction time in s	Simulation time in s
(1,15)	15	105.5091	16.2679
(1,25)	25	202.7613	18.7789
(1,75)	75	664.5100	47.1088
(2,25)	50	415.8458	27.1122
(2,35)	70	555.7065	42.4740
(3,15)	45	351.8655	25.6694
(3,25)	75	615.4011	48.5603
(5,15)	75	571.4564	49.8758

Table 4.3.: PMOR approach: Comparison of the reduction and simulation times based on different pairs (k, r_j) with k sample points and local reduced orders r_j.

invariant with respect to the dimension of the reduced-order subsystems, except for minor deviations. This is obvious for the reduction time, since we use the same truncation tolerance for the different setups and restrict the truncation matrices to the maximum order r afterwards. In case of the timings for the simulation the (almost) invariance is caused by the reuse of the LU decompositions. That is, the main effort in the computation of the solution x_{k+1} at t_{k+1} only has to be performed once for each subsystem. Note that the peaks in the relative errors solely appear at zero-crossings. That is, the numerical computation of the relative errors suffers from the division by numbers close to zero.

For the PMOR ansatz and the associated computation of the local truncation matrices V_j, W_j to the parametric systems in the sample points μ_j, $j = 1, \ldots, k$, the iterative rational Krylov algorithm is applied. The parameter sample points are equidistantly distributed on the guide rails. A convergence tolerance of $1e$-4 and a maximum allowed iteration number of 20 are chosen. Again, the simulation of the resulting global ROMs is computed, using the implicit Euler scheme with the constant time step size $\tau = 10$ s. Table 4.3 together with Figure 4.8 show the different combinations of sample points and local reduced dimensions that are investigated within the numerical computations, as well as the corresponding timings for the reduction and forward simulation. Moreover, the pairs (k, r_j) are chosen such that a global reduced dimension of $r = 75$ is not exceeded. In Figure 4.9, the displacement trajectories in

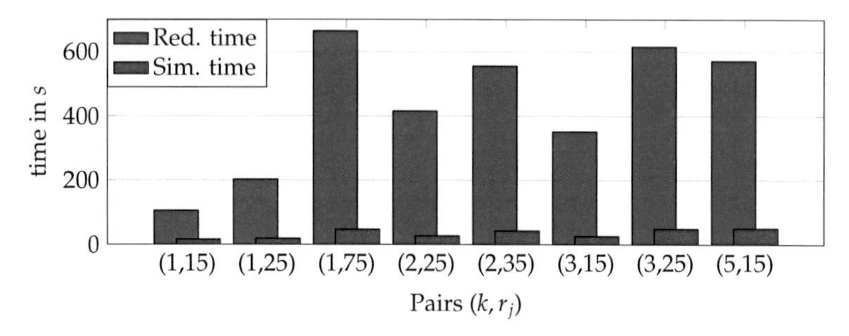

Figure 4.8.: PMOR approach: Comparison of the reduction and simulation times based on different pairs (k, r_j) with k sample points and local reduced orders r_j.

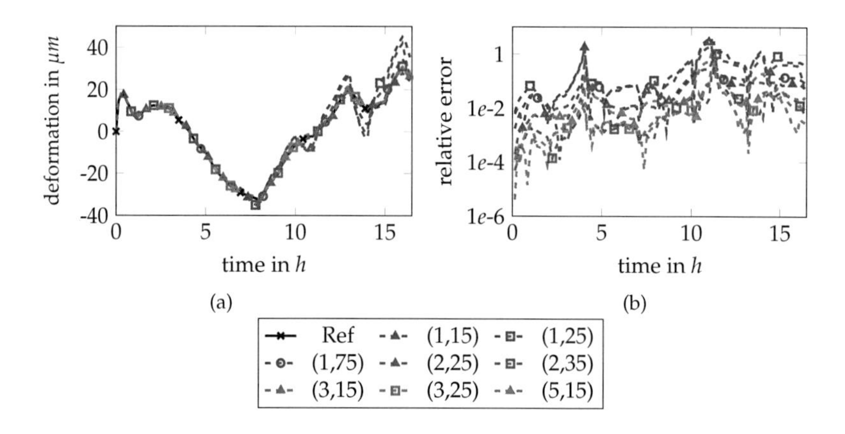

Figure 4.9.: PMOR approach: Displacement (a) and corresponding relative errors (b) in output y_1 for pairs (k, r_j) with k parameter sample points and local reduced-order models of dimension r_j.

the outputs, as well as the corresponding relative errors are depicted for the different combinations of local reduced orders and numbers of sample points. From the results, we observe that increasing the number of sample points, as well as the local reduced order increases the accuracy of the global reduced-order model. Further, the relative errors reveal that, except for the regions of zero-crossings, the reduced-order outputs reach an accuracy in the range of one per thousand if $k \geq 2$ number of samples and a global dimension of $r > 50$ are chosen. Still, it can also be seen that the relative errors slightly increase for t tending to the final time.

Comparing both approaches regarding the same level of accuracy, the PMOR strategy is to prefer with respect to the offline timings. Recall that the offline phase of the PMOR ansatz contains the computation of the truncation matrices and the pre-computation of the reduced-order affine representation. In case of the SLS strategy, the offline part includes the determination of the truncation matrices for all subsystems and the pre-computation of the state transformations $W_{\alpha^+}^T E V_{\alpha^-}$. However, for the presented setup, the SLS ansatz is the method of choice with respect to the online simulation. Since a constant time step is chosen for the forward simulation, the LU-decompositions can be reused and thus accelerates the simulation process compared to the PMOR ansatz. Here, the parameter evaluation of the parameter-varying matrices, based on the pre-computed affine summands, does not allow a repeated utilization of the LU-decompositions within the Euler scheme.

4.3.2. BT for LTV Systems

In order to use the most efficient implementation of the BT procedure for LTV systems for the comparison of the SLS, PMOR, and BT for LTV systems strategies, here we investigate the influence and the performance of a number of time integration methods that are applied to the DLEs. To be more precise, in the following, we consider the BDF(1), BDF(2), Ros(1) and Ros(2) integration schemes, given in Chapter 5. These are based on the efficient LRSIF implementation of the DLE solution methods, being presented in Chapter 6.

Heated Beam:

Consider the 1D heat equation

$$\begin{aligned}
c_p\rho\frac{\partial\Theta}{\partial t}(t,z) &= \kappa\frac{\partial^2\Theta}{\partial z^2}(t,z) + \delta(z-\xi(t))u(t), && (t,z) \in (0,T)\times(0,\ell),\\
\Theta(t,0) &= 0, \qquad \Theta(t,\ell) = 0, && t \in (0,T),\\
\Theta(0,z) &= 0, && z \in (0,\ell),
\end{aligned} \tag{4.38}$$

with a heat source moving along a beam of length ℓ. Here, $\Theta(t,z)$ denotes the temperature distribution at time t and the spatial position z. The material parameters c_p, ρ and κ are the specific heat capacity, the mass density and the heat conductivity, respectively. Furthermore, $\delta(z)$ denotes the Dirac delta function, and $\xi(t)$ and $u(t)$ describe the position of the moving source and the induced heat flux density, respectively, at time t. A finite element discretization of (4.38) with $n+2$ equidistant grid points leads to a system (4.1) with the time-invariant matrices $E, A \in \mathbb{R}^{n\times n}$ and the time-varying input matrix $B(t) \in \mathbb{R}^n$. The output matrix is chosen to be $C(t) = B(t)^T$. That is, the temperature is observed at the location of the heat source. In fact, here we consider a single-input-single-output (SISO) system. Moreover, given $E = E^T$,

Method		BDF(1)	BDF(2)	Ros(1)	Ros(2)
Time in s	DLE	1 396.91	2 242.02	1 448.93	14 383.45
	SRM	0.93	1.04	1.06	2.79

Table 4.4.: BT for LTV systems - Heated beam: the computation times for solving the DLE and the square root method.

$A = A^T$, $B(t) = C(t)^T$ together with the relation $P(t_0) = Q(t_f)$ for the initial condition of the controllability DLE (4.18) and the final condition of the observability DLE (4.19), respectively, we have $P(t) = Q(t_f - t)$ for $t \in [0, t_f]$. Thus, only one DLE has to be solved in Step 1 of Algorithm 4.4. We consider a system dimension of $n = 2\,500$ and the time horizon $t \in [0, 100]\ s$ with a time step size $\tau = 1\ s$. The heat flux, serving as the system input is chosen to be $u(t) = 50\ \frac{W}{m^2}$ and the location of the heat source is defined by $\xi(t) = (\ell t)/T$. In this example, the truncation tolerance for the Hankel singular values is 1e-10. The computation times for the solution of the DLE (4.18), as well as for the application of the square root method are presented in Table 4.4. Figure 4.10 depicts the dimensions of the reduced-order models with respect to the solutions of the DLEs based on the different DLE solvers (a), the Hankel singular values σ_1, σ_3, and σ_6 for all integrators (b) and the decay of the Hankel singular values on the entire time horizon (c), respectively. The system outputs of the full and the reduced-order models, as well as the associated relative errors are depicted in Figures 4.10(d) and 4.10(e), respectively. As expected, the results are independent of the choice of the integration schemes provided that the solutions to the DLEs are of a suitable accuracy. Thus, the use of the most efficient DLE solver, with respect to computation time and storage consumption is recommended. Referring the timings given in Table 4.4, the BDF(1) and Ros(1) schemes show the best performance. For a detailed introduction and investigation of a number of integration schemes, see Chapters 5 and 6.

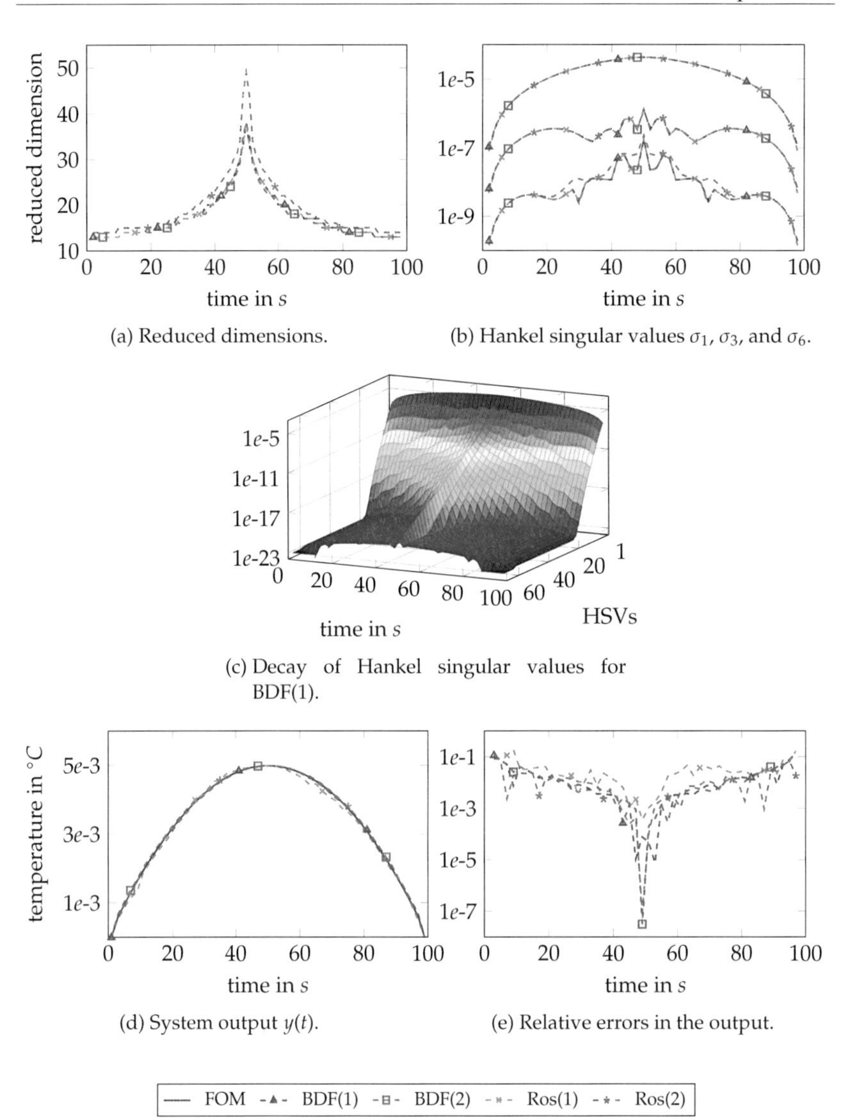

(a) Reduced dimensions.

(b) Hankel singular values σ_1, σ_3, and σ_6.

(c) Decay of Hankel singular values for BDF(1).

(d) System output $y(t)$.

(e) Relative errors in the output.

Figure 4.10.: Heated beam: (a) dimensions of the reduced state, (b) the Hankel singular values σ_1, σ_3, and σ_6 for the BDF methods of order 1, 2, and the Rosenbrock schemes of order 1 and 2, c Hankel singular values for the BDF method of order 1, (d) the outputs of the full and the reduced-order models, and (e) relative errors in the output.

Nonlinear Convection-Diffusion: The Burgers Equations

We consider the 1D Burgers equation

$$\begin{aligned} \dot{x}(t,z) &= \nu \frac{\partial^2}{\partial z^2} x(t,z) - x(t,z)\frac{\partial}{\partial z} x(t,z) + u(t,z), && (t,z) \in (0,T)\times(0,1), \\ x(t,0) &= 0, \quad x(t,1) = 0, && t \in (0,T), \\ x(0,z) &= x_0(z), && z \in (0,1), \end{aligned} \tag{4.39}$$

describing a convection-diffusion problem with the diffusion coefficient ν. The Burgers equation models turbulences, see, e.g., [43, 44], and can be used as a model for shock waves, supersonic flow about airfoils, traffic flows, acoustic transmission and many more. A spatial discretization using finite differences yields a nonlinear model equation. In order to obtain a linear model, a linearization around a pre-computed reference trajectory x_{ref} is applied. The resulting linear time-varying model reads

$$\begin{aligned} E\dot{x}(t) &= A(t)\left(x(t) - x_{ref}(t)\right) + Bu(t) + f(t, x_{ref}(t)) \\ &= A(t)x(t) + Bu(t) + \tilde{f}(t, x_{ref}(t)), \\ y &= Cx(t) \end{aligned} \tag{4.40}$$

with $E = I_n$ and a time-dependence given solely in $A(t)$. For details on the linearization and discretization see, e.g., [100] and the references given therein. We consider a SISO system of dimension $n = 2\,500$. The matrix $B \in \mathbb{R}^n$ is constructed in a way such that the input acts on a grid region of 1 % of the grid nodes around the middle of the spatial domain $(0,1)$. That is, $B_i = 1$ for $i = \lfloor \frac{n}{2} \rfloor \pm \lfloor 0.01n \rfloor$ and zero, otherwise. Here, $\lfloor x \rfloor := \max\{k \in \mathbb{Z} \,|\, x \geq k\}$ denotes the rounding of the argument to the nearest integer value towards zero. The matrix $C \in \mathbb{R}^{1\times n}$ observes the grid point 5 spatial increments to the right of the middle point, i.e., $C_i = 1$ for $i = \lfloor \frac{n}{2} \rfloor + 5$ and zero, otherwise. Note that the inhomogeneity $\tilde{f}(t, x_{ref})$ in (4.40) does not affect the model reduction procedure.

Following a common trick to handle inhomogeneities presented in e.g., [83], system (4.40) can be reformulated into a homogeneous state-space system (4.1) while the system matrices do not change. For that, let $\tilde{x}$ be a solution of the uncontrolled system

$$E\dot{x}(t) = A(t)x(t) + \tilde{f}(t).$$

such that the inhomogeneity is given in the form $\tilde{f}(t) = E\dot{\tilde{x}}(t) - A(t)\tilde{x}(t)$. Then, the state equation in (4.40) becomes

$$\begin{aligned} E\dot{x}(t) &= A(t)x(t) + Bu(t) + E\dot{\tilde{x}}(t) - A(t)\tilde{x}(t) \\ \Leftrightarrow E\left(\dot{x}(t) - \dot{\tilde{x}}(t)\right) &= A(t)\left(x(t) - \tilde{x}(t)\right) + Bu(t). \end{aligned}$$

Finally, defining $z(t) := x(t) - \tilde{x}(t)$, we derive the homogeneous state-space system

$$E\dot{z}(t) = A(t)z(t) + Bu(t), \tag{4.41}$$

for which the BT MOR procedure for LTV systems can be applied. Then, given the truncation matrices $V(t)$, $W(t)$ the corresponding ROM with reduced state $\hat{z}$ reads

$$W(t)^T EV(t)\dot{\hat{z}} = W(t)^T \left(A(t)V(t) - E\dot{V}(t)\right)\hat{z}(t) + W(t)^T Bu(t). \tag{4.42}$$

Now, let $\hat{x}$ and $\hat{\tilde{x}}$ satisfy

$$\begin{aligned} x(t) &= V(t)\hat{x}(t), \quad \dot{x}(t) = V(t)\dot{\hat{x}}(t) + \dot{V}(t)\hat{x}(t), \\ \tilde{x}(t) &= V(t)\hat{\tilde{x}}(t), \quad \dot{\tilde{x}}(t) = V(t)\dot{\hat{\tilde{x}}}(t) + \dot{V}(t)\hat{\tilde{x}}(t), \end{aligned}$$

such that

$$V(t)\hat{z}(t) = z(t) = x(t) - \tilde{x}(t) = V(t)\left(\hat{x}(t) - \hat{\tilde{x}}(t)\right).$$

Then, with $\hat{z}(t) = \hat{x}(t) - \hat{\tilde{x}}(t)$, the ROM (4.42) of the difference system (4.41) reads

$$\begin{aligned} W(t)^T EV(t)\left(\dot{\hat{x}}(t) - \dot{\hat{\tilde{x}}}(t)\right) &= W(t)^T \left(A(t)V(t) - E\dot{V}(t)\right)\left(\hat{x}(t) - \hat{\tilde{x}}(t)\right) + W(t)^T Bu(t) \\ \Leftrightarrow W(t)^T EV(t)\dot{\hat{x}}(t) &= W(t)^T \left(A(t)V(t) - E\dot{V}(t)\right)\hat{x}(t) + W(t)^T Bu(t) \\ &\quad + W(t)^T\Big(E\left(V(t)\dot{\hat{\tilde{x}}}(t) + \dot{V}(t)\hat{\tilde{x}}\right) - A(t)V(t)\hat{\tilde{x}}(t)\Big) \\ &= W(t)^T \left(A(t)V(t) - E\dot{V}(t)\right)\hat{x}(t) + W(t)^T Bu(t) \\ &\quad + W(t)^T\underbrace{\Big(E\dot{\tilde{x}}(t) - A(t)\tilde{x}(t)\Big)}_{=\tilde{f}(t)}, \end{aligned}$$

which in fact represents a reduced-order state equation

$$\hat{E}(t)\dot{\hat{x}}(t) = \hat{A}(t)\hat{x}(t) + \hat{B}(t)u(t) + \hat{f}(t)$$

of the inhomogeneous system (4.40), where $\hat{E}$, $\hat{A}$, $\hat{B}$ are computed as presented in Step 3 of Algorithm 4.4 and the reduced representation of the inhomogeneity $\tilde{f}(t)$ is given by $\hat{f}(t) = W(t)^T \tilde{f}(t)$.

The model is simulated on the time horizon $[0, 3]$ s with time step size $\tau = 2e\text{-}2$ s. The input is chosen to be $u(t) = 2(\sin(2\pi t) + 1)$ for $t \in [0, t_f]$. The truncation tolerance for the Hankel singular values was $1e\text{-}7$. For this example, no explicit time-affine representation of the time-varying system matrix, similar to Equation (4.16), is available. Furthermore, its artificial generation would lead to a considerable large formulation. Thus, the reduced system matrix $\hat{A}(t)$ is computed at every step of the forward simulation. Analogously to the previous example, Table 4.5 shows the computation times for the solution of the DLEs and the square root method. Figure 4.11 depicts the dimensions of the reduced-order models with respect to the DLE solution methods (a), the Hankel singular values σ_1, σ_3, and σ_6 (b), the decay of the Hankel singular values

Method		BDF(1)	BDF(2)	Ros(1)	Ros(2)
Time in s	DLE	4 389.66	3 893.65	4 378.24	17 862.08
	SRM	1.23	1.52	1.35	1.51

Table 4.5.: BT for LTV systems - Burgers equation: the computation times for solving the DLE and the square root method.

over time (c), the output trajectories of the full and the reduced-order models (d), as well as the corresponding relative errors (e). Here, the Ros(1) method results in an erroneous behavior due to the chosen step size and therefore results in an insufficient accuracy for the solution of the DLEs. That is, the Ros(1) scheme requires a smaller step size in order to achieve a compatible accuracy compared to the other methods. This would lead to an increasing computational time and therefore makes the Ros(1) method unsuitable for this example, compared to e.g., the BDF methods.

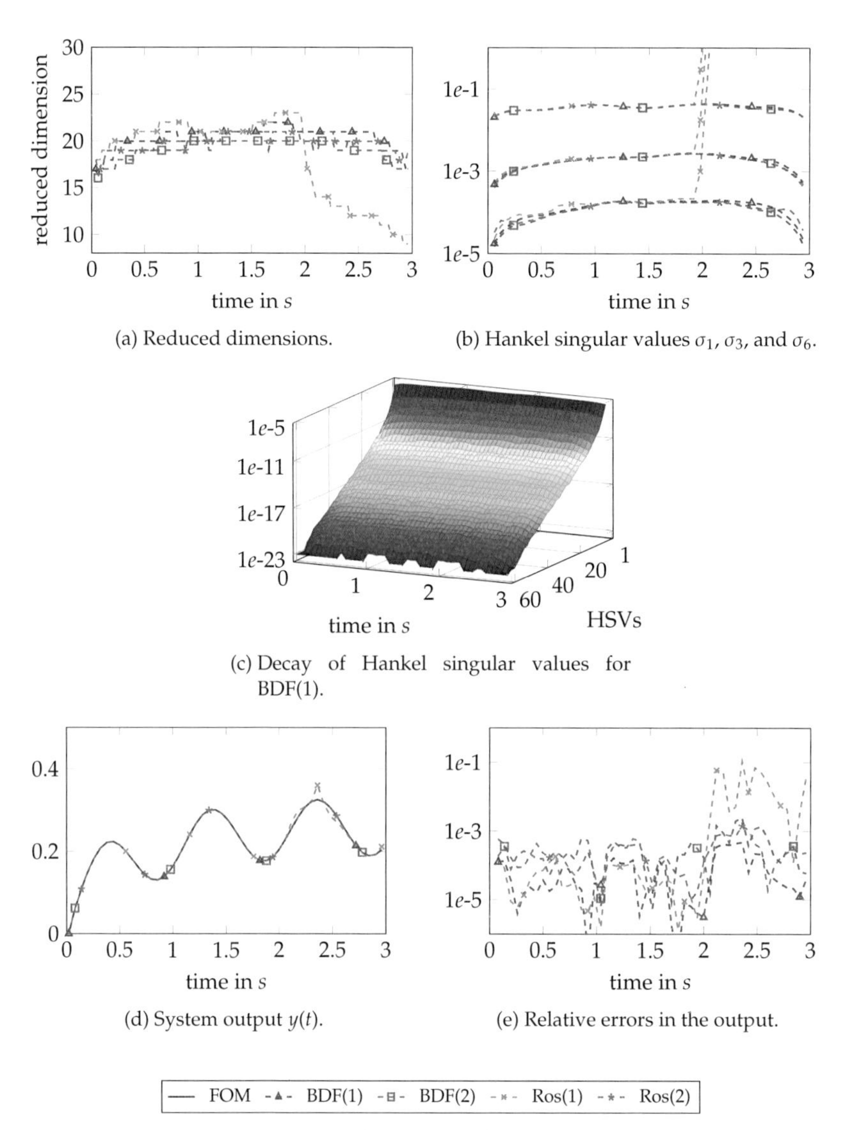

(a) Reduced dimensions.

(b) Hankel singular values σ_1, σ_3, and σ_6.

(c) Decay of Hankel singular values for BDF(1).

(d) System output $y(t)$.

(e) Relative errors in the output.

Figure 4.11.: Burgers equation: (a) dimensions of the reduced state at different times, (b) the Hankel singular values σ_1, σ_3, and σ_6 for the BDF methods of order 1, 2, 3, and 6 and the Rosenbrock schemes of order 1 and 2, (c) Hankel singular values for the BDF method of order 1, (d) the outputs of the full and the reduced-order models, and (e) relative errors in the output.

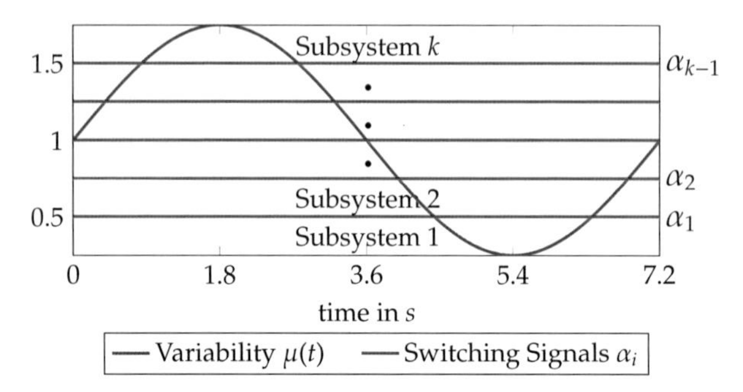

Figure 4.12.: SLS approach: Trajectory of the variability $\mu(t)$ and schematic definition of the SLSs based on the midpoints of the horizontal thresholds.

4.3.3. Time-Varying Rail Example

The rail example is part of the Oberwolfach Model Reduction Benchmark Collection[1] and represents a spatial finite element discretization of a heat transfer problem originating from an optimal cooling problem for steel rail profiles [34, 168]. The semi-discretized model is given in terms of an LTI system (E, A, B, C). In order to obtain an LTV model $(t; E, A, B, C)$, we introduce an artificial time-variability $\mu(t) = \frac{3}{4}\sin(\frac{2\pi t}{t_f}) + 1 \in [0.25, 1.75]$ to the system matrix A. As a result, we obtain an LTV system with constant matrices E, B, C and a time-varying system matrix $A(t) = \mu(t)A$. A grid size, resulting in a state dimension of $n = 1\,357$, is chosen. The system provides the matrices B and C with $m = 7$ inputs and $p = 6$ outputs, respectively. We consider the simulation time interval $[0, t_f]$ s with $t_f = 720$. It should be noted that the time scale of the rail model is internally scaled by the factor 100. That is, the simulation time scale, considered here, corresponds to a real-world time interval of $[0, 7.2]$ s. The timings below, such as time intervals and time step sizes, are given in terms of the real time. Further, a zero initial condition and an input $u(t) = -50$ °C describing the cooling of the steel profile are chosen.

Switched Linear System Formulation

As mentioned before, the basis for the computation of satisfactory MOR results is an appropriate SLS formulation such that the full-order SLS is capable of reconstructing the main features of the LTV model. Therefore, first of all we investigate the number k of subsystems necessary for a moderate approximation of the variability $\mu(t)$. In order to setup the SLSs, the variability $\mu(t)$ is divided into different equidistant horizontal thresholds (see Figure 4.12), whose boundaries α_i serve as the switching instances.

[1]Available from: https://simulation.uni-freiburg.de/downloads/benchmark

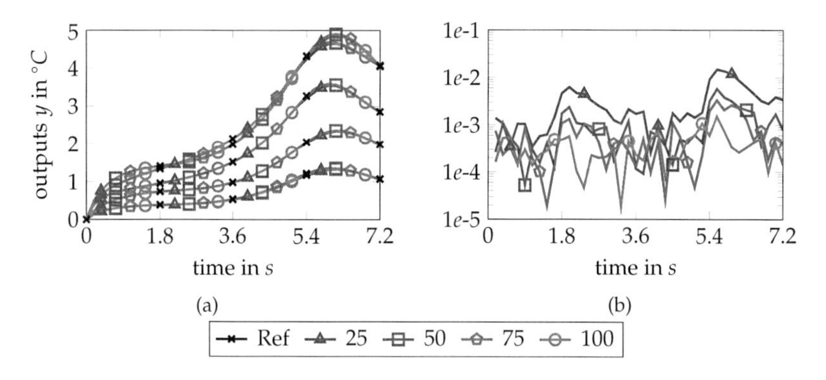

Figure 4.13.: SLS approach: Full-order output trajectories (a) and the relative errors of output y_1 (b) for $k \in \{25, 50, 75, 100\}$ linear subsystems.

That is, the switching signal is defined as

$$\alpha(t) = i, \text{ if } \alpha_{i-1} < \mu(t) \leq \alpha_i$$

with $i \in \mathcal{J} = \{1, \ldots, k\}$ denoting the active subsystem. Then, the midpoints of the intervals $[\alpha_{i-1}, \alpha_i]$ defined by the switching instances define the piecewise constant subsystems such that the variable system matrix is given by

$$A_\alpha = \frac{\alpha_i + \alpha_{i-1}}{2} A.$$

The LTV reference output trajectories and the outputs of the full-order SLSs consisting of $k \in \{25, 50, 75, 100\}$ subsystems are presented in Figure 4.13(a). Moreover, the corresponding relative errors are given in 4.13(b). Since, the several relative errors of the outputs all show the same behavior with respect to the level of accuracy associated to the number of subsystems, as a representative, the relative errors of output y_1 are given in Figure 4.13(b). Note that the relative error of the full-order SLS defines the lower bound for the relative reduction error. That is, it defines the best reduction quality that can be reached. For the upcoming comparison, the representation is restricted to the SLS based on $k = 100$ subsystems that results in a relative error within the range of one per ten thousand except for the time interval $[510, 580]$ s.

Parametric Model Formulation

Given the system matrix in the form $A(t) = \mu(t)A$ in fact directly defines the affine representation

$$A(\mu) = A_0 + f_1(\mu)A_1 + \cdots + f_{m_A}(\mu)A_{m_A}$$

Method		Reduction time in s		Simulation time in s
SLS	$r = 10$	144.0677		0.0262
	$r = 50$	155.5655		0.0450
PMOR	$(1, 25) \Rightarrow r = 25$	4.6896		0.0605
	$(3, 25) \Rightarrow r = 75$	17.0541		0.1133
		DLE	SRM	
BT LTV	tol=1e-2$\Rightarrow r(t) \in [6, 8]$	2657.5974	1.4539	0.1021
	tol=1e-4$\Rightarrow r(t) \in [12, 17]$		1.5432	0.1842

Table 4.6.: Comparison of timings and reduced dimensions for the SLS, PMOR, and BT for LTV systems approach.

for the parameter-varying matrix $A(\mu)$ within the PMOR ansatz with $m_A = 1$, A_0 being the zero matrix, $f_1(\mu) = \mu(t)$ and $A_1 = A$. Admittedly, this is the most comfortable setup possible for the parametric reduction framework. Similar to the previous example, different combinations of sample points and local reduced dimensions are investigated. Similar to the SLS setup, the sample points were chosen on an equidistant grid over the parameter interval.

BT for LTV Systems Formulation

The standard formulation of the BT for LTV systems ansatz does not require a reformulation. In order to efficiently perform the forward simulation of the ROM, we also compute global truncation matrices, similar to the global pair V, W within the PMOR ansatz. Ensuring the stability of the resulting ROM, we set $V = W$. Moreover, the variable system matrix $A(t)$ requires a(n) (time-) affine representation that, for this example, is given by $A(t) = f_1(t)A_1$ with $f_1(t) = \mu(t)$ and $A_1 = A$. Again, this formulation is equivalent to the parameter-affine representation within the PMOR method.

Comparison of the SLS, PMOR, and BT for LTV systems Approaches

The forward simulation of the SLS, PMOR and LTV representation of the rail model is performed, using the implicit Euler scheme based on a constant time integration step size of $\tau = 0.036\,s$, resulting in 200 time steps. The same step size is applied to the DLE solvers within the BT approach for the actual LTV system. As mentioned above, we use the SLS formulation based on $k = 100$ subsystems. Motivated by the results from the heated beam and Burgers equation examples for the BT approach applied to the LTV systems from Section 4.3.2, we use the low-rank symmetric indefinite factorization based BDF(1) scheme in order to compute the reachability and observability Gramians.

Table 4.6 presents the reduction and simulation timings. Exemplarily, the output trajectory y_1 and the associated relative errors are depicted in Figure 4.14. As for

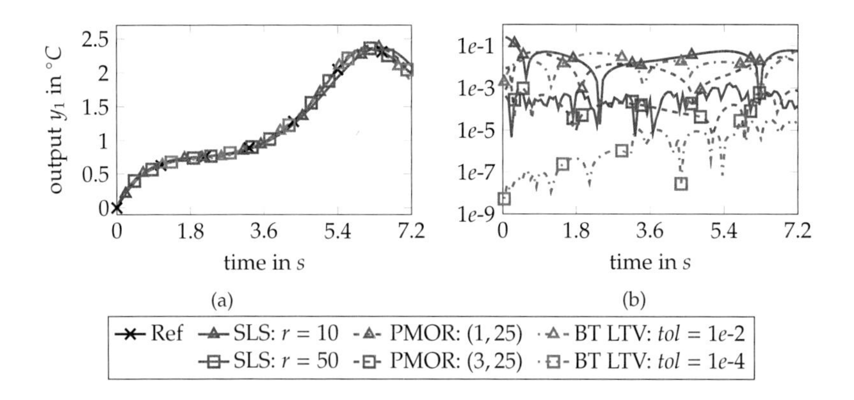

Figure 4.14.: Comparison of the SLS, PMOR and BT LTV approaches: Output trajectory y_1 (a) and the associated relative errors (b).

the previous example the SLS approach yields the best results with respect to the simulation (online) timings, whereas the PMOR procedure is superior regarding the reduction time. The BT method for LTV systems results in the largest offline timings by far that is due to expensive solution of the DLEs. In addition, using time-varying truncation matrices $V(t)$, $W(t)$ within the BT for LTV ansatz requires the computation of the reduced-order system matrix $\hat{A}(t)$ at every time step of the forward simulation of the ROM and therefore yields the largest online timings among the MOR strategies. Still, the computational effort of computing the system Gramians at each discrete time instance pays off concerning the relative errors as the results for a truncation tolerance of 1e-4 reveal. As briefly mentioned in Section 4.2, computing the truncation matrices at each increment of the time discretization used for the solution of the DLEs, in some sense coincides with the determination of the local truncation matrices in the sample points chosen within the PMOR approach. That is, increasing the parameter sample points in the PMOR method allows us to reach the same level of accuracy as for the BT for LTV systems procedure. Nevertheless, the PMOR method offers more flexibility in the choice of sample points, since the BT for LTV scheme is restricted by the choice of the time step size, necessary for the accurate solution of the reachability and observability DLEs. Recall that the accuracy of the reduced-order SLS is restricted by the approximation quality of the full-order SLS. For this example, the SLS based on low-dimensional subsystems of dimension $r = 50$ already reaches the error bound prescribed by the full-order SLS based on $k = 100$ subsystems (see Figure 4.13(b)).

In order to further improve the simulation timings for the BT method for LTV systems, a time-affine reformulation of the variable system matrix $A(t)$ is utilized and globally defined truncation matrices V_{glob}, W_{glob} are computed analogously to the Steps 4 and 5 of Algorithm 4.3 for the PMOR approach. In Figure 4.15 and Table 4.7, we present the

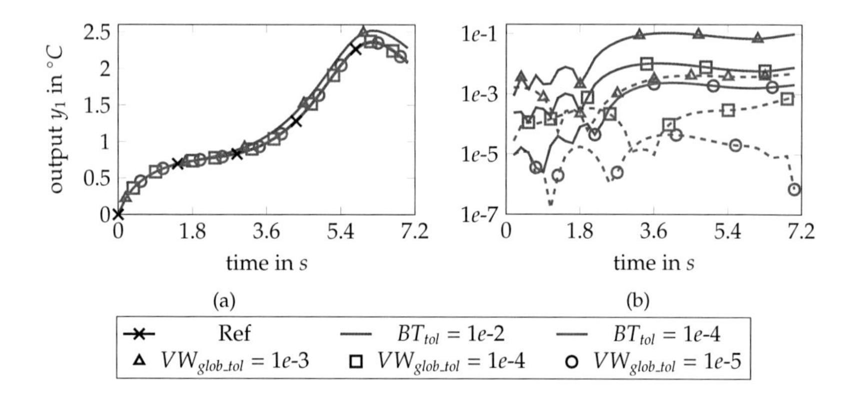

Figure 4.15.: BT LTV with global projection matrices: Output trajectory y_1 (a) and the associated relative errors (b) for different truncation tolerances for the HSVs and projection matrices.

BT_{tol}	Reduction time in s DLE	SRM	glob. V, W	VW_{glob_tol}	Simulation time in s
1e-2	2818.2867	1.3934	4.6609	1e-3 $\Rightarrow r = 38$	0.0230
				1e-4 $\Rightarrow r = 53$	0.0319
				1e-5 $\Rightarrow r = 68$	0.1110
1e-4		1.4252	6.7939	1e-3 $\Rightarrow r = 53$	0.0333
				1e-4 $\Rightarrow r = 68$	0.1187
				1e-5 $\Rightarrow r = 83$	0.1364

Table 4.7.: Comparison of timings and reduced dimensions for the BT for LTV systems approach based on the tolerances BT_{tol}, VW_{glob_tol} for the truncation of the HSVs within BT and the concatenated global truncation matrices, respectively.

results of the MOR strategy with respect to the truncation tolerance BT_{tol} within the BT method, as well as the truncation tolerance VW_{glob_tol} of the concatenated matrices V_{glob}, W_{glob}. Note that this means, the solutions to the DLEs are computed once and afterwards the truncation criterion within the SRM routine based on BT_{tol} and the number of global truncation basis vectors chosen by VW_{glob_tol} are varied. Concerning the online timings for the forward simulation of the ROM it can be observed that the BT method for LTV systems based on global truncation matrices can compete with the SLS and PMOR strategies. Moreover, the obtained relative errors are still superior. That is, neglecting the offline timings, the BT procedure for LTV control systems, based on global truncation matrices, is a noteworthy reduction technique.

4.4. Summary and Conclusion

Three different approaches performing model order reduction for linear time-varying state-space systems are presented. The first two methods rely on the reformulation of the time-dependent system such that the theory of time-invariant systems can be applied. The first of which uses a finite number of LTI subsystems in order to represent the time-variability by a switched linear system. Then, balanced truncation is used in order to separately reduce the subsystems. The single reduced-order submodels are combined by a state transformation at every switching instance during the simulation process. In the second case the actual LTV model equations are expressed in terms of an LPV system, such that the well known parametric MOR techniques based on IRKA can be applied. Finally, the third approach uses the balanced truncation method directly applied to the original LTV system under consideration. Here, the main difficulty is to solve the arising reachability and observability DLEs.

For an efficient implementation of the parametric reduction approach, an affine parameter representation of the LPV system matrices is required. It is further stated that the gain of the entire approach relies on a moderate number of summands of these representations. In contrast to that, the switched linear systems approach shows best performance for a small number of subsystems that of course can only hold for a rather weak time-dependence of the system. In other words, only slow and tightly limited changes in the system matrices will lead to a desirably small number of different subsystems. As mentioned before, the computational effort for the balanced truncation approach for LTV systems is dominated by the solution of the DLEs and thus would benefit from the DLEs to be non-stiff. Moreover, the solution of differential matrix equations for large-scale models is in general a very time and storage consuming process such that, concerning the offline timings, the fields of applications are rather limited at the moment. However, given the solutions to the DLEs, the numerical experiments have proven the LTV BT method, using global projectors, to be comparable with respect to the online timings.

CHAPTER 5

TIME INTEGRATION METHODS FOR DIFFERENTIAL MATRIX EQUATIONS

Contents

The solution of differential Riccati equations has already been studied for more than four decades until now. Thus, in the literature, there is a large variety of solution methods for DREs. The simplest and at the same time most naive approach is the so-called direct integration method based on a vectorization of the matrix-valued differtential equation in order to apply classical ODE solution methods [148, 125]. Obviously, unrolling the matrices of dimension $n \times n$ directly results in the solution of a vector-valued ODE of size n^2 that is unsuitable for large n. Many other methods have been presented in the past [55, 127, 136, 99, 114]. For a variety of reasons, these methods are not further considered in this thesis. See [149] for a detailed overview. Besides the standard ODE integration methods, also other methods have become of importance. In the last few years, splitting methods have been adapted for the application to DREs, see [98, 190, 192]. Very recently, in [124] a structure preserving solution method for large-scale DREs, using Krylov subspace methods, was proposed.

Having large-scale problems in mind, here we focus on matrix-valued versions of classical ODE methods. Due to the fact that in many control applications fast and slow modes are present, the Equations (2.12) and (2.14) are usually fairly stiff. Hence, the numerical solution demands for implicit time integration methods. In the Sections 5.1-5.3, we state a number of classical implicit time integration methods, applied to the matrix-valued differential equations, previously presented in [59, 29, 149, 30, 31]. In Section 5.4, we additionally investigate the rather new class of peer methods [175] in terms of their applicability to differential matrix equations. To be more precise, implicit and linear implicit Rosenbrock-type peer methods are discussed. Moreover, a reformulation of the latter schemes, avoiding numerous applications of the Jacobian, is presented.

For the sake of notation the methods are stated with respect to their application to the Riccati and Lyapunov equations associated to standard state-space systems with $E = I$. Hence, the following explanations restrict to the standard state-space related DME

$$\dot{X}(t) = F(t, X), \qquad X(t_0) = X_0. \tag{5.1}$$

with

$$F(t, X) = A(t)^T X(t) + X(t)A(t) - X(t)S(t)X(t) + W(t) =: \mathcal{R}(t, X), \tag{5.2}$$

in the DRE case and

$$F(t, X) = A(t)X(t) + X(t)A(t)^T + W(t) =: \mathcal{L}(t, X), \tag{5.3}$$

for the DLE. Moreover, in the remainder, we assume the data to be given as $A_k = A(t_k)$, $S_k = S(t_k)$, and $W_k = W(t_k)$, and for the solution to (5.1) we require $X_k \approx X(t_k)$.

5.1. Backward Differentiation Formulas

The backward differentiation formula (BDF) schemes [54] represent the most popular class of linear multistep methods for the solution of stiff problems. For a general overview on linear multistep methods and a detailed description of the BDF methods, see e.g., [95, 96]. The idea, in contrast to one-step integration schemes, is to also use information given from previous integration steps in order to obtain a higher-order reconstruction of the desired solution trajectory. That is, using methods of order $s \geq 2$ requires additional starting values, that can be obtained by any one-step method of comparable order.

The matrix-valued BDF methods applied to the DME (5.1), see [59, 29, 149], yield

$$X_{k+1} = -\sum_{j=1}^{s} \alpha_j X_{k+1-j} + \tau_k \beta F(t_{k+1}, X_{k+1}) \tag{5.4}$$

s	β	α_1	α_2	α_3	α_4	α_5	α_6
1	1	-1					
2	$\frac{2}{3}$	$-\frac{4}{3}$	$\frac{1}{3}$				
3	$\frac{6}{11}$	$-\frac{18}{11}$	$\frac{9}{11}$	$-\frac{2}{11}$			
4	$\frac{12}{25}$	$-\frac{48}{25}$	$\frac{36}{25}$	$-\frac{16}{25}$	$\frac{3}{25}$		
5	$\frac{60}{137}$	$-\frac{300}{137}$	$\frac{300}{137}$	$-\frac{200}{137}$	$\frac{75}{137}$	$-\frac{12}{137}$	
6	$\frac{60}{147}$	$-\frac{360}{147}$	$\frac{450}{147}$	$-\frac{400}{147}$	$\frac{225}{147}$	$-\frac{72}{147}$	$\frac{10}{147}$

Table 5.1.: Coefficients of the BDF s-step methods up to order $s = 6$.

with the coefficients α_j, $j = 1, \ldots, s$, β that determine the s-step BDF formula. The unique set of coefficients up to a BDF order $s = 6$ are given in Table 5.1 (see, e.g., [10]). Note that methods of order $s > 6$ become unstable and will therefore not be considered. Moreover, the BDF schemes are implicit integration methods and therefore, for nonlinear differential equations, such as the DRE (5.1)+(5.2), require the solution of nonlinear algebraic equations at each integration step. Hence, usually a Newton's method is embedded into the integration scheme, see e.g., [10, 58, 95, 96].

BDF for DREs: Applied to the differential Riccati equation (5.1)+(5.2), the BDF scheme results in the Riccati-BDF difference equation

$$-X_{k+1} - \sum_{j=1}^{s} \alpha_j X_{k+1-j} + \tau_k \beta (W_{k+1} + A_{k+1}^T X_{k+1} + X_{k+1} A_{k+1} - X_{k+1} S_{k+1} X_{k+1}) = 0.$$

Combining the constant, linear, and quadratic terms with respect to X_{k+1}, we obtain the formulation

$$\tilde{A}_{k+1}^T X_{k+1} + X_{k+1} \tilde{A}_{k+1} - X_{k+1} \tilde{S}_{k+1} X_{k+1} + \tilde{W}_{k+1} = 0 \tag{5.5}$$

with $\tilde{A}_{k+1} = \tau_k \beta A_{k+1} - \frac{1}{2} I$, $\tilde{S}_{k+1} = \tau_k \beta S_{k+1}$, and $\tilde{W}_{k+1} = \tau_k \beta W_{k+1} - \sum_{j=1}^{s} \alpha_j X_{k+1-j}$. This, in fact, is an algebraic Riccati equation for X_{k+1} that needs to be solved in every time step of the BDF procedure. This can be done using, e.g., Krylov subspace methods or Newton's method. The recent surveys [35, 186] give a detailed overview on a number of ARE solvers. As in the classical textbooks [95, 96], here we restrict to Newton's method, that results in the solution of an algebraic Lyapunov equation in every Newton iteration step, see Section 2.1.3.

BDF for DLEs: Analogously, the BDF scheme, applied to the DLE (5.1)+(5.3), yields the ALE

$$\tilde{A}_{k+1} X_{k+1} + X_{k+1} \tilde{A}_{k+1}^T = -\tilde{W}_{k+1}, \tag{5.6}$$

where $\tilde{A}_{k+1} = \tau_k \beta A_{k+1} - \frac{1}{2} I$ and $\tilde{W}_{k+1} = \tau_k \beta W_{k+1} - \sum_{j=1}^{s} \alpha_j X_{k+1-j}$ does not change with respect to the DRE case. Still, the BDF scheme, applied to the DLE, does not require the application of a nonlinear solver inside the time integration.

Anyway, inside the innermost iteration, both the Newton's method for the solution of the ARE (5.5), as well as the BDF scheme for the DLE end up in the solution of an algebraic Lyapunov equation. The solution of those ALEs using the alternating directions implicit (ADI) iteration or Krylov subspace methods are stated in Section 2.1.4. In the remainder, we use the abbreviations BDF(1), BDF(2) for the BDF methods of order 1 and 2, respectively.

5.2. Rosenbrock Methods

In order to avoid the solution of nonlinear algebraic equations for the solution of (5.1) with nonlinear $F(t, X)$, here we consider the Rosenbrock methods [165]. These methods belong to the class of one-step methods. In particular, the Rosenbrock time integration schemes denote the subclass of linearly implicit one-step methods and can be derived as a linearization of the well-known diagonally implicit Runge-Kutta (DIRK) methods. The idea of the Rosenbrock schemes can be interpreted as the application of one Newton step to each stage of the DIRK methods and directly working with the Jacobian matrix, or an approximation. As a result, only linear systems of algebraic equations need to be solved in every time step of the time integration process. For details on one-step procedures and Runge-Kutta-type methods, we refer to classical textbooks as, e.g., [95, 96]. For the Rosenbrock methods, an in-depth study can be found in [57, Chapter 9]. A general s-stage Rosenbrock scheme, applied to the matrix-valued DME (5.1), [149, 30], reads

$$
\begin{aligned}
X_{k+1} &= X_k + \tau_k \sum_{i=1}^{s} b_i K_i, \\
K_i &= F\left(t_{k,i}, X_k + \tau_k \sum_{j=1}^{i-1} a_{i,j} K_j\right) + \tau_k J_k \sum_{j=1}^{i} \gamma_{i,j} K_j, \ i = 1, \ldots, s.
\end{aligned}
\tag{5.7}
$$

In the literature, many authors start their explanations for autonomous ODE systems, see, e.g., [96, Section IV.7]. Then a general Rosenbrock scheme for non-autonomous ODEs is based on an autonomization and thus yields

$$
\begin{aligned}
X_{k+1} &= X_k + \tau_k \sum_{i=1}^{s} b_i K_i, \\
K_i &= F\left(t_{k,i}, X_k + \tau_k \sum_{j=1}^{i-1} a_{i,j} K_j\right) + \tau_k J_k \sum_{j=1}^{i} \gamma_{i,j} K_j + \gamma_i \tau_k^2 F_{t_k}, \ i = 1, \ldots, s.
\end{aligned}
\tag{5.8}
$$

In both representations, we have $t_{k,i} = t_k + c_i\tau_k$, $i = 1, \dots, s$, and $b_i, \gamma_{i,j}, a_{i,j}, c_i = \sum_{j=1}^{i-1} a_{i,j}$, and $\gamma_i = \sum_{j=1}^{i-1} \gamma_{i,j}$ are the method's coefficients. Further, J_k denotes the Jacobian and the matrix $K_i \in \mathbb{R}^{n\times n}$ represents the solution of the ith-stage of the method. In the latter scheme, defined for non-autonomous ODEs being autonomized, time t is added to the dependent variables and therefore the additional variable F_{t_k}, denoting the time derivative $\frac{\partial F}{\partial t}(t_k, X(t_k))$ of F at $(t_k, X(t_k))$, appears in the procedure. This term may be disadvantageous for the stability of the method, see [57, Chapter 9.5] for details. Therefore, in the remainder, we neglect the time derivative F_{t_k} and restrict our considerations to the scheme (5.7) defined for non-autonomous ODEs, not being autonomized.

Finally, simple algebraic manipulations yield the integration schemes

$$
\begin{aligned}
X_{k+1} &= X_k + \tau_k \sum_{i=1}^{s} b_i K_i, \\
(I - \tau_k \gamma_{i,i} J_k) K_i &= F\left(t_{k,i}, X_k + \tau_k \sum_{j=1}^{i-1} a_{i,j} K_j\right) + \tau_k J_k \sum_{j=1}^{i-1} \gamma_{i,j} K_j, \; i = 1, \dots, s.
\end{aligned}
\tag{5.9}
$$

In practice, methods with $\gamma_{11} = \cdots = \gamma_{ss} = \gamma$ are often of special interest. That is, the systems to be solved do not change for the several stages at a certain time t_k and therefore, for an efficient implementation, parts of the solver can be reused.

The Jacobian J_k for the above stated Rosenbrock representations is given by the Fréchet derivative $\frac{\partial F}{\partial X}(t_k, X_k)$ of F at X_k.

Rosenbrock method for DREs: For the DRE (5.1)+(5.2) the Jacobian is given by the shifted Lyapunov operator

$$
J_k := \frac{\partial \mathcal{R}}{\partial X}(t_k, X_k) : U \to (A_k - S_k X_k)^T U + U(A_k - S_k X_k). \tag{5.10}
$$

Applying the specific Jacobian to the left hand side of the integration scheme (5.9), we obtain

$$
\begin{aligned}
X_{k+1} &= X_k + \tau_k \sum_{i=1}^{s} b_i K_i, \\
\tilde{A}_k^T K_i + K_i \tilde{A}_k &= -\mathcal{R}\Big(t_{k,i}, X_k + \tau_k \sum_{j=1}^{i-1} \alpha_{i,j} K_j\Big) - \tau_k J_k \sum_{j=1}^{i-1} \gamma_{i,j} K_j,
\end{aligned}
\tag{5.11}
$$

with $\tilde{A}_k = \tau_k \gamma (A_k - S_k X_k) - \frac{1}{2} I$.

Rosenbrock method for DLEs: In case of the DLE (5.1)+(5.3), which again is linear, the Fréchet derivative $\mathcal{L}(t_k, X_k)$ is the Lyapunov operator itself and reads

$$J_k := \frac{\partial \mathcal{L}}{\partial X}(t_k, X_k) : U \to A_k U + U A_k^T. \tag{5.12}$$

Then, similar to Equation (5.11), the Rosenbrock scheme becomes

$$\begin{aligned} X_{k+1} &= X_k + \tau_k \sum_{i=1}^{s} b_i K_i, \\ \tilde{A}_k K_i + K_i \tilde{A}_k^T &= -\mathcal{L}\Big(t_{k,i}, X_k + \tau_k \sum_{j=1}^{i-1} \alpha_{i,j} K_j\Big) - \tau_k J_k \sum_{j=1}^{i-1} \gamma_{i,j} K_j \end{aligned} \tag{5.13}$$

with $\tilde{A}_k = \tau_k \gamma A_k - \frac{1}{2} I$.

That in fact means that one algebraic Lyapunov equation needs to be solved in each stage of every time step, for both, the DRE and DLE. Moreover, given the representation (5.7), at stage i the application of the Jacobian J_k to all previously computed stages K_j, $j = 1, \ldots, i-1$ is required. Therefore, in practice usually a set of auxiliary variables, avoiding the application of J_k, is introduced. For details, see standard textbooks, as e.g. [96, Chapter IV.7].

As representatives, we consider a 1-stage Rosenbrock scheme of first-order, as well as a 2-stage integrator of second-order. For $s = 1$ and the coefficients $b_1 = 1$, $\gamma = 1$ and $c_1 = 0$, the first-order Rosenbrock method (Ros(1)) applied to the DRE reads

$$\begin{aligned} X_{k+1} &= X_k + \tau_k K_1, \\ \tilde{A}_k^T K_1 + K_1 \tilde{A}_k &= -\mathcal{R}(t_k, X_k) \end{aligned} \tag{5.14}$$

with $\tilde{A} = \tau_k (A_k - S_k X_k) - \frac{1}{2} I$ and

$$\begin{aligned} X_{k+1} &= X_k + \tau_k K_1, \\ \tilde{A}_k K_1 + K_1 \tilde{A}_k^T &= -\mathcal{L}(t_k, X_k) \end{aligned} \tag{5.15}$$

with $\tilde{A} = \tau_k A_k - \frac{1}{2} I$ in the DLE case. The Ros(1) scheme in fact represents the linearly implicit Euler method. Further, substituting the solution update K_1 with $K_1 = \frac{X_{k+1} - X_k}{\tau_k}$ and following the reformulations given in [149, 30], the Ros(1) scheme simplifies to

$$\hat{A}_k^T X_{k+1} + X_{k+1} \hat{A}_k = -W_k - X_k S_k X_k - \frac{1}{\tau_k} X_k \tag{5.16}$$

with $\hat{A}_k = A_k - S_k X_k - \frac{1}{2\tau_k} I$ for the DRE and

$$\hat{A}_k X_{k+1} + X_{k+1} \hat{A}_k^T = -W_k - \frac{1}{\tau_k} X_k \tag{5.17}$$

with $\hat{A}_k = A_k - \frac{1}{2\tau_k}I$ for the DLE. Note that this allows to directly compute the sought for solution approximation X_{k+1} at t_{k+1}.

Further, we consider a second-order scheme (Ros(2)) that was, for scalar problems, originally presented in [57, Example 9.1.1] and is further discussed in [36]. In the latter, the second-order method is applied to autonomous atmospheric dispersion problems describing photochemistry, advective, and turbulent diffusive transport. Moreover note that the Rosenbrock formulation, avoiding the application of the Jacobian to the stages K_j, $j = 1, \dots, i-1$, is used.

Applied to the DRE, we obtain

$$\begin{aligned} X_{k+1} &= X_k + \frac{3}{2}\tau_k K_1 + \frac{1}{2}\tau_k K_2, \\ \tilde{A}_k^T K_1 + K_1 \tilde{A}_k &= -\mathcal{R}(t_k, X_k), \\ \tilde{A}_k^T K_2 + K_2 \tilde{A}_k &= -\mathcal{R}(t_{k+1}, X_k + \tau_k K_1) + 2K_1 \end{aligned} \tag{5.18}$$

with $t_{k+1} = t_k + \tau_k$ and $\tilde{A}_k$ as in (5.11), whereas the application to the DLE yields

$$\begin{aligned} X_{k+1} &= X_k + \frac{3}{2}\tau_k K_1 + \frac{1}{2}\tau_k K_2, \\ \tilde{A}_k K_1 + K_1 \tilde{A}_k^T &= -\mathcal{L}(t_k, X_k), \\ \tilde{A}_k K_2 + K_2 \tilde{A}_k^T &= -\mathcal{L}(t_{k+1}, X_k + \tau_k K_1) + 2K_1 \end{aligned} \tag{5.19}$$

with $\tilde{A}_k$ from (5.13).

In Section 6.4.1 a fourth-order scheme (Ros4) with four stages will be used for the computation of a dense reference solution. Details on the Ros4 method and the corresponding coefficients can be found in the Appendix A.

It should be mentioned that for autonomous systems, the Rosenbrock methods of order $s \geq 2$ can be further simplified. For the development of an efficient Ros(2) method for large-scale problems, this is done in Chapter 6.

5.3. Other Implicit Methods

Among a number of implicit integration methods, present in the literature, here we discuss the application of two of the most well-known integrators, namely the Midpoint and Trapezoidal rule, previously introduced in [59].

5.3.1. Midpoint Rule

The Midpoint rule (Mid) applied to the DME (5.1), yields

$$X_{k+1} = X_k + \tau_k F\Big(t_k + \frac{\tau_k}{2}, \frac{1}{2}(X_{k+1} + X_k)\Big).$$

Similar to the BDF scheme for the DRE case an ARE of the form

$$\tilde{A}_{k'}^T X_{k+1} + X_{k+1}\tilde{A}_{k'} - \frac{\tau_k}{4} X_{k+1} S_{k'} X_{k+1} + \tilde{W}_{k'} = 0 \tag{5.20}$$

with $\tilde{W}_{k'} = X_k + \tau_k \left(W_{k'} + \frac{1}{2}\left(A_{k'}^T X_k + X_k A_{k'} - \frac{X_k S_{k'} X_k}{2} \right) \right)$ and $\tilde{A}_{k'} = \frac{1}{2}\left(\tau_k (A_{k'} - \frac{1}{2} S_{k'} X_k) - I \right)$ has to be solved for X_{k+1} at every step k of the integration method. In case of the DLE the resulting ALE is given by

$$\tilde{A}_{k'} X_{k+1} + X_{k+1} \tilde{A}_{k'}^T = -\tilde{W}_{k'}, \tag{5.21}$$

where $\tilde{W}_{k'} = \tau_k W_{k'} + X_k + \frac{\tau_k}{2}\left(A_{k'} X_k + X_k A_{k'}^T \right)$ and $\tilde{A}_{k'} = \frac{1}{2}(\tau_k A_{k'} - I)$. Further, $A_{k'} = A(t_k + \frac{\tau_k}{2})$, $W_{k'} = W(t_k + \frac{\tau_k}{2})$, $S_{k'} = S(t_k + \frac{\tau_k}{2})$.

5.3.2. Trapezoidal Rule

Applying the Trapezoidal rule (Trap) to the DME (5.1), we obtain

$$X_{k+1} = X_k + \frac{\tau_k}{2}(F(t_k, X_k) + F(t_{k+1}, X_{k+1})).$$

Then again, for the DRE this results in the solution of an ARE

$$\tilde{A}_{k+1}^T X_{k+1} + X_{k+1}\tilde{A}_{k+1} - \frac{\tau_k}{2} X_{k+1} S_{k+1} X_{k+1} + \tilde{W}_{k+1} = 0 \tag{5.22}$$

with $\tilde{W}_{k+1} = \frac{\tau_k}{2} W_{k+1} + X_k + \frac{\tau_k}{2}\left(W_k + A_k^T X_k + X_k A_k - X_k S_k X_k \right)$ and $\tilde{A}_{k+1} = \frac{\tau_k}{2} A_{k+1} - \frac{1}{2} I$ and the DLE case yields the ALE

$$\tilde{A}_{k+1} X_{k+1} + X_{k+1} \tilde{A}_{k+1}^T = -\tilde{W}_{k+1} \tag{5.23}$$

with $\tilde{W}_{k+1} = \frac{\tau_k}{2} W_{k+1} + X_k + \frac{\tau_k}{2}\left(W_k + A_k X_k + X_k A_k^T \right)$ and $\tilde{A}_{k+1} = \frac{\tau_k}{2} A_{k+1} - \frac{1}{2} I$.

For the solution of the AREs (5.20) and (5.22), we again refer to Section 2.1.3 and the solution strategies for the ALEs (5.21) and (5.23) are given in Section 2.1.4.

5.4. Peer Methods

Linear multistep and one-step methods, such as the BDF and the Rosenbrock methods, respectively, have been known for many decades and in addition have proven their effectiveness over the years for a wide range of problems. These two traditional classes of time integration methods have always been studied separately. As a unifying framework for investigating stability, consistency and convergence for a wide variety of methods, also containing the aforementioned classes, in [45] the general linear methods (GLMs) were introduced. A more detailed overview on GLMs is, e.g., given in the surveys [46, 47]. Most of the classical methods contain a number of "master" variables $X_{k+1,j}$ and in addition compute a separate number of "slave" variables related to function evaluations $F(\tilde{t}_k, \tilde{X}_k)$ at $t_k \leq \tilde{t}_k \leq t_{k+1}$, $\tilde{X}_k \approx X(\tilde{t}_k)$ that are designated to improve the accuracy and stability properties of the masters. In particular, usually only one master variable for the approximation of the solution in each time interval is employed. Moreover, for different time intervals, as, e.g., for variable time step sizes, these master variables may have distinguished accuracy and stability properties. Now, the idea of the so-called *peer methods* is to define an integration scheme that only contains peer variables, all representing approximations to the exact solution to (5.1) at different time locations, and additionally share the same accuracy and stability properties. That is, no explicit slave variables will be introduced. The class of peer methods first appeared in [175] in terms of linearly implicit integration schemes with peer variables, suitable for parallel computations by only using information from the previous time interval. A number of specific peer schemes and applications are presented in, e.g., [162, 163, 176, 177]. Further, for a recent detailed overview see [193, Chapters 5,10].

5.4.1. Implicit Peer Methods

A general implicit peer method, applied to the matrix-valued initial value problem (5.1) reads

$$X_{k,i} = \sum_{j=1}^{s} b_{i,j} X_{k-1,j} + \tau_k \sum_{j=1}^{i} g_{i,j} F(t_{k,j}, X_{k,j}). \tag{5.24}$$

The convergence order of these methods is restricted to $s-1$. Thus, additionally using function values from the previous time interval, two-step peer methods of order s can be constructed. Under special conditions even a superconvergent subclass of the implicit peer methods with convergence order $s+1$ can be found. Details on the above convergence statements are given in [187]. The corresponding scheme becomes

$$X_{k,i} = \sum_{j=1}^{s} b_{i,j} X_{k-1,j} + \tau_k \sum_{j=1}^{s} a_{i,j} F(t_{k-1,j}, X_{k-1,j}) + \tau_k \sum_{j=1}^{i} g_{i,j} F(t_{k,j}, X_{k,j}). \tag{5.25}$$

Here, s is the number of stages and

$$t_{k,j} = t_k + c_j \tau_k, \tag{5.26}$$

where the c_j, $j = 1, \ldots, s$ with $c_s = 1$, $t_{k,s} = t_{k+1}$ define the locations of the peer variables for the time step $t_k \to t_{k+1}$. In general, $c_j < 0$ for some j will also be allowed. However, the following schemes will only use $c_j > 0$. The variables $b_{i,j}$, $a_{i,j}$, and $g_{i,j}$ are the determining coefficients of the method. For the same reasons as for the Rosenbrock schemes (Section 5.2), methods with $g_{i,i} = \gamma$, $i = 1, \ldots, s$, are of particular interest. Note that, given from the order conditions, the coefficients will in general depend on the step size ratio τ_k/τ_{k-1} of two consecutive time steps. Moreover, the computation of the coefficients is based on highly sophisticated optimization processes. Therefore, for details on the order conditions and the computation of the associated coefficients, we refer to [187] and the references therein. However, the variables $X_{k,i}$ denote approximations to the exact solution $X(t_{k,i}) = X(t_k + c_i\tau_k)$ of (5.1). It is easy to verify that the scheme (5.24) is recovered from (5.25) by simply setting $a_{i,j} = 0$ and therefore in the remainder the statements are restricted to the more general class (5.25) of implicit peer methods.

The implicit peer scheme (5.25) applied to the DRE (5.2) with $F(t_{k,i}, X_{k,i}) = \mathcal{R}(t_{k,i}, X_{k,i})$ reads

$$\tilde{A}_{k,i}^T X_{k,i} + X_{k,i}\tilde{A}_{k,i} - X_{k,i}\tilde{S}_{k,i}X_{k,i} + \tilde{W}_{k,i} = 0, \qquad i = 1, \ldots, s, \tag{5.27}$$

where

$$\begin{aligned}
\tilde{A}_{k,i} &= \tau_k g_{i,i} A_{k,i} - \frac{1}{2}I, \quad \tilde{S}_{k,i} = \tau_k g_{i,i} S_{k,i}, \\
\tilde{W}_{k,i} &= \tau_k g_{i,i} W_{k,i} + \sum_{j=1}^{s} b_{i,j} X_{k-1,j} + \tau_k \sum_{j=1}^{s} a_{i,j}\mathcal{R}(t_{k-1,j}, X_{k-1,j}) + \tau_k \sum_{j=1}^{i-1} g_{i,j}\mathcal{R}(t_{k,j}, X_{k,j}).
\end{aligned}$$

Moreover, we have $A_{k,i} = A(t_{k,i})$, $W_{k,i} := W(t_{k,i})$ and $S_{k,i} = S(t_{k,i})$ with $t_{k,i}$ given by Equation (5.26). Finally, from Equation (5.27), we have that, according to the number of peer variables to be computed, s AREs have to be solved at every time step of the peer method.

The linear DLEs represented by $F(t_{k,i}, X_{k,i}) = \mathcal{L}(t_{k,i}, X_{k,i})$ from (5.3) yield the peer scheme

$$\tilde{A}_{k,i}X_{k,i} + X_{k,i}\tilde{A}_{k,i}^T = -\tilde{W}_{k,i}, \; i = 1, \ldots, s \tag{5.28}$$

with

$$\begin{aligned}
\tilde{A}_{k,i} &= \tau_k g_{i,i} A_{k,i} - \frac{1}{2}I, \\
\tilde{W}_{k,i} &= \tau_k g_{i,i} W_{k,i} + \sum_{j=1}^{s} b_{i,j} X_{k-1,j} + \tau_k \sum_{j=1}^{s} a_{i,j}\mathcal{L}(t_{k-1,j}, X_{k-1,j}) + \tau_k \sum_{j=1}^{i-1} g_{i,j}\mathcal{L}(t_{k,j}, X_{k,j}).
\end{aligned}$$

Similar to the DRE, s ALEs have to be solved at every time step of the implicit peer method (5.25) applied to the DLE.

Recall that the BDF methods, the Midpoint, and Trapezoidal rules require the solution of one ARE/ALE at every time step. Therefore, regarding the use of direct solution

c_1 :	0.4831632475943920	c_2 :	1.0000000000000000
$b_{1,1}$:	−0.3045407685048590	$b_{1,2}$:	1.3045407685048591
$b_{2,1}$:	−0.3045407685048590	$b_{2,2}$:	1.3045407685048591
$g_{1,1}$:	0.2584183762028040	$g_{1,2}$:	0.0000000000000000
$g_{2,1}$:	0.4376001712448750	$g_{2,2}$:	0.2584183762028040

Table 5.2.: 2-stage implicit peer method of order 2.

methods for the occurring matrix equations and a certain fixed time step size τ, the expected computational effort of the peer methods is in general s-times higher than that of the other implicit methods. Still, from the fact that s peer variables with the same accuracy and stability properties are computed within every time interval, the peer methods allow us to use larger step sizes in order to achieve a comparable accuracy. A comparison and detailed investigation is given in Section 6.4.

For the numerical experiments, we consider an implicit peer method of order 1 with the coefficients $c_1 = 1$, $b_{1,1} = 1$ and $g_{1,1} = 1$ and a method of order 2. The coefficients of the 2-stage implicit peer method, given in Table 5.2, were provided by the group of Prof. R. Weiner[1] at the Martin-Luther-Universität Halle. For these methods, the number of stages and the convergence order coincide. In what remains, the implicit peer methods of order 1, 2 are denoted by Peer(1) and Peer(2), respectively.

5.4.2. Rosenbrock-Type Peer Methods

Similar to the idea of the classical Rosenbrock methods (Section 5.2) the solution of nonlinear matrix equations can be avoided. Therefore, we also consider the linearly implicit peer methods in terms of the two-step Rosenbrock-type peer schemes

$$\begin{aligned}(I - \tau_k g_{i,i} J_k) X_{k,i} = \sum_{j=1}^{s} b_{i,j} X_{k-1,j} + \tau_k \sum_{j=1}^{s} a_{i,j} \left(F(t_{k-1,j}, X_{k-1,j}) - J_k X_{k-1,j} \right) \\ + \tau_k J_k \sum_{j=1}^{i-1} g_{i,j} X_{k,j},\end{aligned} \tag{5.29}$$

introduced in [162]. Therein, the authors consider methods with $g_{11} = \cdots = g_{ss} = \gamma$. For the comprehensive derivation of coefficients $a_{i,j}$, $b_{i,j}$ and $g_{i,j}$ that result in stable schemes (5.29), for arbitrary step size ratios, we refer to [162, Section 3]. As for the classical Rosenbrock methods, J_k denotes the Jacobian represented by the Fréchet derivative of F at (t_k, X_k). That is, replacing the Jacobian J_k in (5.29) by (5.10), for the

[1] http://www.mathematik.uni-halle.de/wissenschaftliches_rechnen/ruediger_weiner/

solution of the DRE, the procedure reads

$$
\begin{aligned}
\tilde{A}_{k,i}^T X_{k,i} + X_{k,i}\tilde{A}_{k,i} &= -\tilde{W}_{k,i} \quad i = 1, \dots, s, \\
\tilde{W}_{k,i} &= \sum_{j=1}^{s} b_{i,j} X_{k-1,j} + \tau_k \sum_{j=1}^{s} a_{i,j} \left(\mathcal{R}(t_{k-1,j}, X_{k-1,j}) - (\hat{A}_k^T X_{k-1,j} + X_{k-1,j}\hat{A}_k) \right) \\
&\quad + \tau_k \sum_{j=1}^{i-1} g_{i,j} (\hat{A}_k^T X_{k,j} + X_{k,j}\hat{A}_k)
\end{aligned}
\tag{5.30}
$$

with the matrices $\hat{A}_k = A_k - S_k X_k$ and $\tilde{A}_{k,i} = \tau_k g_{i,i}\hat{A}_k - \frac{1}{2}I$.

Using the Jacobian (5.12), the application of (5.29) to the DLE yields

$$
\begin{aligned}
\tilde{A}_{k,i} X_{k,i} + X_{k,i}\tilde{A}_{k,i}^T &= -\tilde{W}_{k,i} \quad i = 1, \dots, s, \\
\tilde{W}_{k,i} &= \sum_{j=1}^{s} b_{i,j} X_{k-1,j} + \tau_k \sum_{j=1}^{s} a_{i,j} \left(\mathcal{L}(t_{k-1,j}, X_{k-1,j}) - (A_k X_{k-1,j} + X_{k-1,j} A_k^T) \right) \\
&\quad + \tau_k \sum_{j=1}^{i-1} g_{i,j} (A_k X_{k,j} + X_{k,j} A_k^T),
\end{aligned}
\tag{5.31}
$$

where $\tilde{A}_{k,i} = \tau_k g_{i,i} A_k - \frac{1}{2}I$. In its current appearance, the scheme (5.29) involves the solution of an ALE at each stage and further requires the application of the Jacobian J_k to the sums $\sum_{j=1}^{s} a_{i,j} X_{k-1,j}$ and $\sum_{j=1}^{i} g_{i,j} X_{k,j}$ of the previous and current solution approximations, respectively. In order to at least avoid the application of the Jacobians to the sum of new variables $X_{k,j}$, a reformulation, similar to what is standard for the classical Rosenbrock schemes, see, e.g., [96, Chapter IV.7], based on the variables

$$
Y_{k,i} = \sum_{j=1}^{i} g_{i,j} X_{k,j}, \; i = 1, \dots, s \Leftrightarrow \mathbf{Y}_k = (G \otimes I_n)\mathbf{X}_k
\tag{5.32}
$$

is proposed. Here, $\mathbf{X}_k = (X_{k,i})_{i=1}^s$, $\mathbf{Y}_k = (Y_{k,i})_{i=1}^s \in \mathbb{R}^{sn \times n}$ and $\otimes$ denotes the Kronecker product. Provided that $g_{i,i} \neq 0$, $\forall i$, the lower triangular matrix $G = (g_{i,j})$ is nonsingular and the original variables $X_{k,i}$ can be recovered from the relation

$$
\mathbf{X}_k = (G^{-1} \otimes I_n)\mathbf{Y}_k \Leftrightarrow X_{k,i} = \sum_{j=1}^{i} \mathbf{g}_{i,j} Y_{k,j}, \; i = 1, \dots, s
\tag{5.33}
$$

where $G^{-1} = (\mathbf{g}_{i,j})$ and $\mathbf{g}_{i,i} = \frac{1}{g_{i,i}}$. Moreover, we have

$$
\sum_{j=1}^{s} a_{i,j} J_k X_{k-1,j} = J_k \sum_{j=1}^{s} a_{i,j} X_{k-1,j}
$$

and from (5.33), we obtain

$$
\begin{aligned}
\sum_{j=1}^{s} a_{i,j} X_{k-1,j},\ i = 1,\dots,s \Leftrightarrow \quad & ((a_{i,j}) \otimes I)\mathbf{X}_k \\
&= ((a_{i,j}) \otimes I)(G^{-1} \otimes I)\mathbf{Y}_k \\
&= ((a_{i,j})G^{-1} \otimes I)\mathbf{Y}_k \qquad \Leftrightarrow \sum_{j=1}^{s} \mathbf{a}_{i,j} Y_{k-1,j},\ i = 1,\dots,s
\end{aligned}
\tag{5.34}
$$

with the coefficients

$$
(\mathbf{a}_{i,j}) = (a_{i,j})G^{-1}. \tag{5.35}
$$

Analogously, for the sum $\sum_{j=1}^{s} b_{i,j} X_{k-1,j}$, we have

$$
\sum_{j=1}^{s} b_{i,j} X_{k-1,j} = \sum_{j=1}^{s} \mathbf{b}_{i,j} Y_{k-1,j},\ i = 1,\dots,s
$$

with $(\mathbf{b}_{i,j}) = (b_{i,j})G^{-1}$. Then, inserting the auxiliary variables (5.32) into (5.29) and dividing the result by τ_k, the linearly implicit scheme reformulates to

$$
\begin{aligned}
\left(\frac{1}{\tau_k g_{i,i}} I - J_k\right) Y_{k,i} = &\sum_{j=1}^{s} \frac{\mathbf{b}_{i,j}}{\tau_k} Y_{k-1,j} + \sum_{j=1}^{s} a_{i,j} F(t_{k-1,j}, \sum_{\ell=1}^{j} \mathbf{g}_{j,\ell} Y_{k-1,\ell}) \\
&- J_k \sum_{j=1}^{s} \mathbf{a}_{i,j} Y_{k-1,j} - \sum_{j=1}^{i-1} \frac{\mathbf{g}_{i,j}}{\tau_k} Y_{k,j}, \quad i = 1,\dots,s.
\end{aligned}
\tag{5.36}
$$

Again, replacing the Jacobian J_k by (5.10), the modified Rosenbrock-type scheme, applied to the DRE, reads

$$
\begin{aligned}
&\tilde{A}_{k,i}^T Y_{k,i} + Y_{k,i} \tilde{A}_{k,i} = -\tilde{W}_{k,i}, \\
&\tilde{W}_{k,i} = \sum_{j=1}^{s} \frac{\mathbf{b}_{i,j}}{\tau_k} Y_{k-1,j} + \sum_{j=1}^{s} a_{i,j} \mathcal{R}(t_{k-1,j}, \sum_{\ell=1}^{j} \mathbf{g}_{j,\ell} Y_{k-1,\ell}) \\
&\qquad - \sum_{j=1}^{s} \mathbf{a}_{i,j} (\hat{A}_k^T Y_{k-1,j} + Y_{k-1,j} \hat{A}_k) - \sum_{j=1}^{i-1} \frac{\mathbf{g}_{i,j}}{\tau_k} Y_{k,j}
\end{aligned}
\tag{5.37}
$$

with $\hat{A}_k$ from the original scheme and $\tilde{A}_{k,i} = \hat{A}_k - \frac{1}{2\tau_k g_{i,i}} I$.

Using the Jacobian from (5.12), the application to the DLE yields

$$\begin{aligned}
\tilde{A}_{k,i} Y_{k,i} + Y_{k,i} \tilde{A}_{k,i}^T &= -\tilde{W}_{k,i}, \\
\tilde{W}_{k,i} &= \sum_{j=1}^{s} \frac{\mathbf{b}_{i,j}}{\tau_k} Y_{k-1,j} + \sum_{j=1}^{s} a_{i,j} \mathcal{L}(t_{k-1,j}, \sum_{\ell=1}^{j} \mathbf{g}_{j,\ell} Y_{k-1,\ell}) \\
&\quad - \sum_{j=1}^{s} \mathbf{a}_{i,j} (A_k Y_{k-1,j} + Y_{k-1,j} A_k^T) - \sum_{j=1}^{i-1} \frac{\mathbf{g}_{i,j}}{\tau_k} Y_{k,j},
\end{aligned} \tag{5.38}$$

where $\tilde{A}_{k,i} = A_k - \frac{1}{2\tau_k g_{i,i}} I$.

Recall that the introduction of the auxiliary variables is capable of avoiding the application of the Jacobian J_k to the sum of current stage variables. Still, the application remains for the sum of the previously determined peer variables. Moreover, in contrast to the one-step Rosenbrock methods, the reformulation in terms of the auxiliary variables requires the reconstruction of the original solution approximations and thus doubles the online storage amount for storing the solution and auxiliary trajectory during the runtime of the integration method. Further, note that, in general, the function F, defining the ODE under consideration, has to be applied to the expression $\sum_{\ell=1}^{j} \mathbf{g}_{j,\ell} Y_{k-1,\ell}$. For small-scale problems, this results in a number of additional matrix-matrix additions, that is, a cheap computation, compared to the solution of the ALEs. Still, in the case of large-scale applications and the required low-rank strategies, the size of the right hand side factors of the ALEs will dramatically blow up and might therefore not be applicable in practice. Thus, for the implementation of these methods, we use a mixed formulation with respect to the variables $X_{k,j}$ and $Y_{k,j}$. For details see Sections 6.1.5 and 6.2.5. Furthermore, for linear functions F, such as the Lyapunov operator $\mathcal{L}$, the function evaluations can efficiently be computed, using reformulations similar to (5.34). For more details and an in-depth investigation, see Section 6.1.5 of the following chapter.

Summarizing, the linearly implicit peer methods result in the solution of s ALEs, just like the classical Rosenbrock methods, but directly compute the sought for solutions, instead of additional stage variables. In particular, the stage variables K_i from the Rosenbrock methods in Section 5.2 are of low stage order and therefore the integration procedures may suffer from order reduction. The computation of peer variables in the Rosenbrock-type peer scheme, sharing the same accuracy and stability properties, can overcome this well-known disadvantage and again allows us to use larger time steps compared to the classical one-step Rosenbrock methods.

For scalar and vector-valued ODEs, the peer methods are in general competitive to classical multi-step and one-step methods. Still, a number of function evaluations $\sum_{j=1}^{s} a_{i,j} F(t_{k-1,j}, X_{k-1,j})$ at the previously determined solutions have to be computed that, in the case of matrix-valued differential equations, results in a considerable computational effort. In Section 6.4, we show the performance of the peer methods applied to matrix-valued differential equations.

c_1 :	0.500000000000000	c_2 :	1.000000000000000
$a_{1,1}$:	−0.292893218813452	$a_{1,2}$:	0.585786437626905
$a_{2,1}$:	−1.085786437626905	$a_{2,2}$:	1.878679656440358
$b_{1,1}$:	−0.414213562373095	$b_{1,2}$:	1.414213562373095
$b_{2,1}$:	−0.414213562373095	$b_{2,2}$:	1.414213562373095
$g_{1,1}$:	0.292893218813452	$g_{1,2}$:	0.000000000000000
$g_{2,1}$:	0.500000000000000	$g_{2,2}$:	0.292893218813452

Table 5.3.: 2-stage Rosenbrock-type peer method of order 2.

As for the BDF and implicit peer methods, the numerical experiments, presented in Section 6.4, are restricted to Rosenbrock-type peer methods of order 1 and 2. These, we denote by RosPeer(1) and RosPeer(2), respectively, and analogously for the modified versions, avoiding the application of the Jacobian to the currently computed solution approximations, we use the abbreviations mRosPeer(1) and mRosPeer(2). The RosPeer(1) method is given by the coefficients $c_1 = 1$, $a_{1,1} = 1$, $b_{1,1} = 1$ and $g_{1,1} = 1$. The coefficients of the RosPeer(2) scheme are listed in Table 5.3.

CHAPTER 6

EFFICIENT SOLUTION OF LARGE-SCALE DIFFERENTIAL MATRIX EQUATIONS

Contents

In Chapter 5, the basic formulations of a number of implicit solution methods for ordinary differential equations and their application to differential Riccati and Lyapunov equations, associated to control systems in standard state-space form, are presented. For small-scale problems these methods can easily be implemented using direct solvers for the solution of the arising algebraic Lyapunov equations, as mentioned in Section 2.1.4. Anyway, for sparse and large-scale DMEs it is crucial to find an efficient solution strategy in order to solve the ALEs. Here, the low-rank Lyapunov solvers, described in Section 2.1.4, are considered. The performance explicitly depends on the chosen factorization of the corresponding right hand sides and hence of the solutions to the ALEs. In Section 6.1, we review the classical low-rank Cholesky-type factorizations, that, related to standard DREs, were presented in [149, 30, 31] for the BDF and Rosenbrock methods. Further, we show that for methods of order $s \geq 2$, the classical low-rank formulations have to deal with complex data and arithmetic. Then, Section 6.2 presents the symmetric indefinite factorizations, avoiding the aforementioned drawbacks. Finding efficient solution strategies, further, requires the elimination of redundant information in the accumulating solution factors. This is performed by factorization dependent, rank-revealing compression techniques, presented in Section 6.3.

However, in order to avoid the inversion of $E \neq I$ for generalized state-space systems (E, A, B, C) in a final step, the integration methods presented for standard systems (Chapter 5) are re-transformed. Motivated by the application in Chapter 3, we consider the optimal control related GDRE

$$E^T\dot{X}(t)E = A(t)^T X(t)E + E^T X(t)A(t) + W(t) - E^T X(t)S(t)X(t)E = \mathcal{R}(t, X(t)) \tag{6.1}$$

with the matrices $S(t) = B(t)R^{-1}(t)B(t)^T$, $W(t) = C(t)^T Q(t)C(t)$, representing the reverse-in-time version of Equation (3.24). As a representative for the GDLEs, we consider the reachability Lyapunov equation

$$E(t)\dot{X}(t)E(t)^T = A(t)X(t)E(t)^T + E(t)X(t)A(t)^T + W(t) = \mathcal{L}(t, X(t)) \tag{6.2}$$

from the balanced truncation framework, given in Equation (4.18) of Section 4.2, with $W(t) = B(t)B(t)^T$. The observability Lyapunov equation can be solved analogously by replacing $A(t)$ with $A(t)^T$ and $B(t)$ with $C(t)^T$. Further, recall that the input and output matrices $B(t) \in \mathbb{R}^{n \times m}$ and $C(t) \in \mathbb{R}^{q \times n}$, respectively, are assumed to be thin, rectangular matrices with $m,\ q \ll n$ and therefore the solutions to the GDRE (6.1) and the GDLE (6.2) are observed to be of low numerical rank and thus can be well approximated by a product of low-rank factors.

In what remains, we will go through the integration schemes applied to the GDRE (6.1) in detail and briefly state the analogue results for the representative GDLE (6.2). Moreover, for some of the methods, the integration schemes can be significantly simplified considering the autonomous case, i.e., $\mathcal{R}(t, X) = \mathcal{R}(X)$ or $\mathcal{L}(t, X) = \mathcal{L}(X)$ and $E,\ A,\ B,\ C$ being constant. For those methods, both the non-autonomous and autonomous representations are investigated in detail.

6.1. Classical Low-Rank Factorization

As already mentioned in Section 2.1.4, the solution of an algebraic Lyapunov equation can often be observed to be of low numerical column rank, if this also applies to the corresponding right hand side. Therefore, in this section, we try to find a product GG^T with G being of low numerical column rank n_G, representing the right hand sides of the algebraic Lyapunov equations of the form

$$A^T XE + E^T XA = -GG^T, \tag{6.3}$$

arising in the innermost iteration of the time integration schemes, presented in Chapter 5. That is, in the remainder, we assume the solutions to the GDRE (6.1) and GDLE (6.2) to be of low numerical rank for all t and the approximate solution $X_j \approx X(t_j)$ admits a low-rank representation $X_j = Z_j Z_j^T$ with $Z_j \in \mathbb{R}^{n \times n_{Z_j}}$, $n_{Z_j} \ll n$ for all $t_j \in [t_0, t_f]$.

6.1.1. Backward Differentiation Formulas

Classical low-rank BDF scheme for GDREs For the BDF methods, applied to the GDRE (6.1), we have to solve the GARE

$$\tilde{A}_{k+1}^T X_{k+1} E + E^T X_{k+1} \tilde{A}_{k+1} - E^T X_{k+1} \tilde{S}_{k+1} X_{k+1} E + \tilde{W}_{k+1} = 0 \tag{6.4}$$

with

$$\tilde{A}_{k+1} = \tau_k \beta A_{k+1} - \frac{1}{2} E, \quad \tilde{S}_{k+1} = \tau_k \beta B_{k+1} B_{k+1}^T,$$
$$\tilde{W}_{k+1} = \tau_k \beta C_{k+1}^T C_{k+1} - E^T \left(\sum_{j=1}^{s} \alpha_j X_{k+1-j} \right) E,$$

that, applying Newton's method, results in the solution of the GALE

$$\begin{aligned} \hat{A}_{k+1}^{(\ell)^T} X_{k+1}^{(\ell)} E + E^T X_{k+1}^{(\ell)} \hat{A}_{k+1}^{(\ell)} &= - \left(\tau_k \beta C_{k+1}^T C_{k+1} - E^T \left(\sum_{j=1}^{s} \alpha_j X_{k+1-j} \right) E \right) \\ &\quad - \tau_k \beta E^T X_{k+1}^{(\ell-1)} B_{k+1} B_{k+1}^T X_{k+1}^{(\ell-1)} E \end{aligned} \tag{6.5}$$

with $\hat{A}_{k+1}^{(\ell)} = \tau_k \beta \left(A_{k+1} - B_{k+1} B_{k+1}^T X_{k+1}^{(\ell-1)} E \right) - \frac{1}{2} E$, for $X_{k+1}^{(\ell)}$ in the ℓ-th Newton step. If now, the solutions to the previous time steps admit a low-rank factorization, then the GALE (6.5) with factorized right hand side can be written in the form

$$\hat{A}_{k+1}^{(\ell)^T} X_{k+1}^{(\ell)} E + E^T X_{k+1}^{(\ell)} \hat{A}_{k+1}^{(\ell)} = -G_k^{(\ell)} G_k^{(\ell)^T} \tag{6.6}$$

with

$$G_k^{(\ell)} = \left[\sqrt{\tau_k \beta} C_{k+1}^T, \ \sqrt{-\alpha_1} E^T Z_k, \ldots, \sqrt{-\alpha_s} E^T Z_{k+1-s}, \ \sqrt{\tau_k \beta} E^T Z_{k+1}^{(\ell-1)} Z_{k+1}^{(\ell-1)^T} B_{k+1} \right] \tag{6.7}$$

of column size $n_{G_k} = q + \sum_{j=1}^{s} n_{Z_{k+1-j}} + m$. Since we never want to form a product $Z_k Z_k^T$ explicitly, the final expression in the right hand side (6.7) should be computed, exploiting the small matrix products in the form $E^T\big(Z_{k+1}^{(\ell-1)}\big(Z_{k+1}^{(\ell-1)^T} B_{k+1}\big)\big)$.

Classical low-rank BDF scheme for GDLEs The GDLE (6.2), plugged into the BDF scheme, yields the GALE

$$\tilde{A}_{k+1} X_{k+1} E^T + E X_{k+1} \tilde{A}_{k+1} = -G_k G_k^T, \tag{6.8}$$

where $\tilde{A}_{k+1} = \tau_k \beta A_{k+1} - \frac{1}{2} E$ and the right hand side factor becomes

$$G_k = \Big[\sqrt{\tau_k \beta} B_{k+1},\ \sqrt{-\alpha_1} E Z_k, \ldots,\ \sqrt{-\alpha_s} E Z_{k+1-s}\Big] \in \mathbb{C}^{n \times m + (\sum_{j=1}^{s} n_{Z_{k+1-j}})}.$$

Due to the alternating signs of the BDF coefficients α_j, $j = 1, \ldots, s$, for order $s \geq 2$, the non-factorized right hand side becomes indefinite and thus, although the actual right hand sides are real and symmetric, yields complex factors $G_k^{(\ell)}$ and G_k for both, the GDRE and GDLE, respectively. That is, the GALEs in the innermost iteration have to be solved with complex arithmetic and the solution demands for complex storage.

Note that the GALEs are linear and therefore the solutions $X_{k+1}^{(\ell)}$ and X_{k+1} of (6.6) and (6.8), respectively could, in general, be split in the form $X = \tilde{X} - \hat{X}$, such that the expressions with differing signs can be separated. That is, defining $X_{k+1} = \tilde{X}_{k+1} - \hat{X}_{k+1}$ for, e.g., the solution of (6.8), we end up with the two GALEs

$$\begin{aligned} \tilde{A}_{k+1} \tilde{X}_{k+1} E^T + E \tilde{X}_{k+1} \tilde{A}_{k+1}^T &= -N_k N_k^T, \\ \tilde{A}_{k+1} \hat{X}_{k+1} E^T + E \hat{X}_{k+1} \tilde{A}_{k+1}^T &= -U_k U_k^T, \end{aligned} \tag{6.9}$$

where $G_k G_k^T = N_k N_k - U_k U_k$ with N_k containing those blocks in G_k with positive sign and accordingly U_k consists of the blocks with negative sign. That is, avoiding complex arithmetic, one additional GALE would have to be solved in every step of the innermost iteration. Still, in general, the split GALEs (6.9) can be solved efficiently, noting that both equations share the same coefficient matrix $\tilde{A}_{k+1}$. Nevertheless, numerous numerical experiments have shown that the superposition of the actual sought for solution X_{k+1} is corrupted by round-off errors and numerical cancellation effects, accumulating over time. In particular, this splitting resulted in failing algorithms for all test runs. Therefore, for the solution of the GALEs, arising in the BDF methods, we restrict to the simple two-term decomposition, resulting in complex factors. Moreover, in the following subsections, we will recognize that all integration methods of order $s \geq 2$ have to deal with complex arithmetic due to the classical low-rank decomposition.

Note that the classical low-rank ADI and Krylov subspace based algorithms applied to Lyapunov equations with real symmetric right hand sides, presented in Section 2.1.4, originally are developed for positive (semi)-definite symmetric or Hermitian right hand sides, given in terms of the factorizations GG^T or GG^H, respectively. Still, there is another scenario to be considered, originating from, e.g., time integration methods of order $s \geq 2$, as we have seen in the BDF example above. There, we need to investigate the applicability of the ADI and Krylov subspace solution methods for real symmetric but indefinite right hand sides $W = GG^T$ with $G \in \mathbb{C}^{n \times n_G}$. This is postponed to the end of this section. In addition, a specific column compression technique, capable of handling the complex data, is presented in Section 6.3.1.

6.1.2. Rosenbrock Methods

Considering the s-stage Rosenbrock methods (5.7), the solution of a nonlinear system of equations is avoided and the scheme for the GDRE, as well as the GDLE, results in the solution of one GALE for each stage of the procedure.

Classical low-rank Rosenbrock methods for GDREs Starting with the linearly implicit Euler method, i.e., the simplified 1-stage Rosenbrock method (5.16), applied to the GDRE (6.1), we have

$$\hat{A}_k^T X_{k+1} E + E^T X_{k+1} \hat{A}_k = -C_k^T C_k - E^T X_k B_k B_k^T X_k E - \frac{1}{\tau_k} E^T X_k E \tag{6.10}$$

with $\hat{A}_k = A_k - B_k B_k^T X_k E - \frac{1}{2\tau_k} E$. Then, the solution factor Z_{k+1} can be achieved, using Algorithm 2.1 and the right hand side factor

$$G_k = \left[C_k^T,\ E^T Z_k (Z_k^T B_k),\ \sqrt{\tfrac{1}{\tau_k}} E^T Z_k \right] \in \mathbb{R}^{n \times (q+m+n_{Z_k})}.$$

The Ros(2) scheme (5.18), applied to the GDRE (6.1), reads

$$X_{k+1} = X_k + \frac{3}{2}\tau_k K_1 + \frac{1}{2}\tau_k K_2, \tag{6.11a}$$

$$\tilde{A}_k^T K_1 E + E^T K_1 \tilde{A}_k = -\mathcal{R}(t_k, X_k), \tag{6.11b}$$

$$\tilde{A}_k^T K_2 E + E^T K_2 \tilde{A}_k = -\mathcal{R}(t_{k+1}, X_k + \tau_k K_1) + 2E^T K_1 E, \tag{6.11c}$$

where $\tilde{A}_k = \tau_k \gamma (A_k - B_k B_k^T X_k E) - \frac{1}{2} E$. In order to find a factorization of the Riccati operator

$$\mathcal{R}(t_k, X_k) = C_k^T C_k + A_k^T X_k E + E^T X_k A_k - E^T X_k B_k B_k^T X_k E, \tag{6.12}$$

with $X_k = Z_k Z_k^T$, first we consider the factorization

$$A_k^T Z_k Z_k^T E + E^T Z_k Z_k^T A_k = \left(A_k^T Z_k + E^T Z_k\right)\left(A_k^T Z_k + E^T Z_k\right)^T - A_k^T Z_k Z_k^T A_k - E^T Z_k Z_k^T E \tag{6.13}$$

of the linear term. Consequently, the right hand side of the first-stage Equation (6.11b) can be factored in the form $\mathcal{R}(t_k, X_k) = G_k^{(1)} G_k^{(1)^T}$, where

$$G_k^{(1)} = \left[C_k^T,\ A_k^T Z_k + E^T Z_k,\ \jmath A_k^T Z_k,\ \jmath E^T Z_k,\ \jmath E^T Z_k (Z_k^T B_k)\right] \in \mathbb{C}^{n\times(q+3n_{Z_k}+m)}.$$

For the second-stage (6.11c), we have

$$\begin{aligned}\mathcal{R}(t_{k+1}, X_k + \tau_k K_1) =& C_{k+1}^T C_{k+1} + A_{k+1}^T X_k E + E^T X_k A_{k+1} - E^T X_k B_{k+1} B_{k+1}^T X_k E \\ &+ \tau_k \left(A_{k+1}^T K_1 E + E^T K_1 A_{k+1}\right) - \tau_k^2 E^T K_1 B_{k+1} B_{k+1}^T K_1 E \\ &- \tau_k \left(E^T X_k B_{k+1} B_{k+1}^T K_1 E + E^T K_1 B_{k+1} B_{k+1}^T X_k E\right)\end{aligned} \tag{6.14}$$

and assuming the solution of (6.11b) to be given as $K_1 = T_1 T_1^T$, $T_1 \in \mathbb{R}^{n\times n_{T_1}}$, the factorization $G_k^{(2)} G_k^{(2)^T}$ of the right hand side reads

$$\begin{aligned}G_k^{(2)} = \Big[& C_{k+1}^T,\ A_{k+1}^T Z_k + E^T Z_k,\ \jmath A_{k+1}^T Z_k,\ \jmath E^T Z_k,\ \jmath E^T Z_k (Z_k^T B_{k+1}),\ \sqrt{\tau_k}(A_{k+1}^T T_1 + E^T T_1), \\ & \jmath \sqrt{\tau_k} A_{k+1}^T T_1,\ \jmath \sqrt{\tau_k} E^T T_1,\ \jmath \sqrt{\tau_k}(E^T T_1 (T_1^T B_{k+1}) + E^T Z_k (Z_k^T B_{k+1})), \\ & \sqrt{\tau_k} E^T T_1 (T_1^T B_{k+1}),\ \sqrt{\tau_k} E^T Z_k (Z_k^T B_{k+1}),\ \jmath \tau_k E^T T_1 (T_1^T B_{k+1}),\ \jmath \sqrt{2} E^T T_1\Big],\end{aligned}$$

where $G_k^{(2)} \in \mathbb{C}^{n\times(q+3n_{Z_k}+4n_{T_1}+5m)}$.

Given an autonomous GDRE, we note that

$$\begin{aligned}\mathcal{R}(X_k + \tau_k K_1) &= C^T C + A^T (X_k + \tau_k K_1) E + E^T (X_k + \tau_k K_1) A - E^T (X_k + \tau_k K_1) B B^T (X_k + \tau_k K_1) E \\ &= \mathcal{R}(X_k) + \tau_k (A - B B^T X_k E)^T K_1 E + \tau_k E^T K_1 (A - B B^T X_k E) - \tau_k^2 E^T K_1 B B^T K_1 E \\ &= \mathcal{R}(X_k) + \tau_k (A - B B^T X_k E - \frac{1}{2\gamma\tau_k} E)^T K_1 E + \tau_k E^T K_1 (A - B B^T X_k E - \frac{1}{2\gamma\tau_k} E) \\ &\quad - \tau_k^2 E^T K_1 B B^T K_1 E + \frac{1}{\gamma} E^T K_1 E \\ &= (1 - \frac{1}{\gamma}) \mathcal{R}(X_k) - \tau_k^2 E^T K_1 B B^T K_1 E + \frac{1}{\gamma} E^T K_1 E\end{aligned}$$

and the Ros(2) scheme (6.11) can be simplified to

$$\begin{aligned}X_{k+1} &= X_k + (2 - \frac{1}{2\gamma}) \tau_k K_1 - \frac{1}{2} \tau_k K_{21}, \\ \tilde{A}_k^T K_1 E + E^T K_1 \tilde{A}_k &= -\mathcal{R}(X_k) = -G_k^{(1)} G_k^{(1)^T}, \\ \tilde{A}_k^T K_{21} E + E^T K_{21} \tilde{A}_k &= -\tau_k^2 E^T K_1 B B^T K_1 E - (2 - \frac{1}{\gamma}) E^T K_1 E = -G_k^{(2)} G_k^{(2)^T},\end{aligned} \tag{6.15}$$

where the right hand side factorizations are given in the form

$$\begin{aligned}G_k^{(1)} &= \left[C^T,\ A^T Z_k + E^T Z_k,\ \jmath A^T Z_k,\ \jmath E^T Z_k,\ \jmath E^T Z_k (Z_k^T B)\right] \in \mathbb{C}^{n\times(q+3n_{Z_k}+m)}, \\ G_k^{(21)} &= \left[\tau E^T T_1 (T_1^T B),\ \sqrt{2 - \tfrac{1}{\gamma}} E^T T_1\right] \in \mathbb{R}^{n\times(m+n_{T_1})}.\end{aligned}$$

Formally, the solution of the second stage can be reconstructed by $K_2 = -K_{21} + (1 - \frac{1}{\gamma})K_1$.

Again, as for the BDF methods, for order $s \geq 2$ some of the right hand sides of the stage GALEs become indefinite and therefore the associated factors are complex

Classical low-rank Rosenbrock methods for GDLEs Analogously to the GDRE case, the first-order Rosenbrock method (5.16), applied to the GDLE (6.2), reads

$$\hat{A}_k X_{k+1} E^T + E X_{k+1} \hat{A}_k^T = -B_k B_k^T - \frac{1}{\tau_k} E X_k E^T \tag{6.16}$$

with $\hat{A}_k = A_k - \frac{1}{2\tau_k} E$, where the right hand side is factorized in the form

$$G_k = \left[B_k, \ \sqrt{\tfrac{1}{\tau_k}} E Z_k \right] \in \mathbb{R}^{n \times (m + n_{Z_k})}.$$

The second-order scheme (5.19) yields,

$$\begin{aligned} X_{k+1} &= X_k + \frac{3}{2}\tau_k K_1 + \frac{1}{2}\tau_k K_2, \\ \tilde{A}_k K_1 E^T + E K_1 \tilde{A}_k^T &= -\mathcal{L}(t_k, X_k) = -G_k^{(1)} {G_k^{(1)}}^T, \\ \tilde{A}_k^T K_2 E^T + E K_2 \tilde{A}_k^T &= -\mathcal{L}(t_{k+1}, X_k + \tau_k K_1) + 2 E K_1 E^T = -G_k^{(2)} {G_k^{(2)}}^T, \end{aligned} \tag{6.17}$$

with

$$\begin{aligned} \tilde{A}_k &= \tau_k \gamma A_k - \frac{1}{2} E, \\ G_k^{(1)} &= \left[B_k, \ A_k Z_k + E Z_k, \ \jmath A_k Z_k, \ \jmath E Z_k, \right] \in \mathbb{C}^{n \times (m + 3 n_{Z_k})}, \\ G_k^{(2)} &= \left[B_{k+1}, \ A_{k+1} Z_k + E Z_k, \ \jmath A_{k+1} Z_k, \ \jmath E Z_k, \right. \\ &\quad \left. \sqrt{\tau_k}(A_{k+1} T_1 + E T_1), \ \jmath \sqrt{\tau_k} A_{k+1} T_1, \ \jmath \sqrt{\tau_k} E T_1, \ \jmath \sqrt{2} E T_1 \right] \in \mathbb{C}^{n \times (m + 3 n_{Z_k} + 4 n_{T_1})}, \end{aligned}$$

for non-autonomous GDLEs. In the autonomous case, the scheme is given by

$$\begin{aligned} X_{k+1} &= X_k + (2 - \frac{1}{2\gamma})\tau_k K_1 - \frac{1}{2}\tau_k K_{21}, \\ \tilde{A}_k K_1 E^T + E K_1 \tilde{A}_k^T &= -\mathcal{L}(X_k) = G_k^{(1)} {G_k^{(1)}}^T, \\ \tilde{A}_k K_{21} E^T + E K_{21} \tilde{A}_k^T &= -(2 - \frac{1}{\gamma}) E K_1 E^T = G_k^{(2)} {G_k^{(2)}}^T, \end{aligned} \tag{6.18}$$

where the right hand sides are factored in the form

$$G_k^{(1)} = \left[B, \ A Z_k + E Z_k, \ \jmath A Z_k, \ \jmath E Z_k, \right] \in \mathbb{C}^{n \times (m + 3 n_{Z_k})}, \quad G_k^{(2)} = \sqrt{2 - \frac{1}{\gamma}} E T_1 \in \mathbb{R}^{n \times n_{T_1}}.$$

6.1.3. Midpoint Rule

Classical low-rank Midpoint rule for GDREs Similar to the BDF methods, the GARE arising within the Midpoint rule reads

$$\tilde{A}_{k'}^T X_{k+1} E + E^T X_{k+1} \tilde{A}_{k'} - E^T X_{k+1} \tilde{S}_{k'} X_{k+1} E + \tilde{W}_{k'} = 0 \tag{6.19}$$

with

$$\begin{aligned}\tilde{A}_{k'} &= \frac{\tau_k}{2}(A_{k'} - \frac{1}{2} B_{k'} B_{k'}^T X_k E) - \frac{1}{2} E, \quad \tilde{S}_{k'} = \frac{\tau_k}{4} B_{k'} B_{k'}^T, \\ \tilde{W}_{k'} &= E^T X_k E + \tau_k C_{k'}^T C_{k'} + \frac{\tau_k}{2}\left(A_{k'}^T X_k E + E^T X_k A_{k'}\right) - \frac{\tau_k}{4} E^T X_k B_{k'} B_{k'}^T X_k E.\end{aligned}$$

Then, applying Newton's method, we obtain the GALE

$$\begin{aligned}\hat{A}_{k'}^{(\ell)^T} X_{k+1}^{(\ell)} E + E^T X_{k+1}^{(\ell)} \hat{A}_{k'}^{(\ell)} &= -\left(\tau_k C_{k'}^T C_{k'} + \check{A}_{k'}^T X_k E + E^T X_k \check{A}_{k'} - \frac{\tau_k}{4} E^T X_k B_{k'} B_{k'}^T X_k E\right) \\ &\quad - \frac{\tau_k}{4} E^T X_{k+1}^{(\ell-1)} B_{k'} B_{k'}^T X_{k+1}^{(\ell-1)} E = -G_k^{(\ell)} G_k^{(\ell)^T}\end{aligned} \tag{6.20}$$

with

$$\hat{A}_{k'}^{(\ell)} = \frac{\tau_k}{2}(A_{k'} - \frac{1}{2} B_{k'} B_{k'}^T (X_k + X_{k+1}^{(\ell-1)}) E) - \frac{1}{2} E, \quad \check{A}_{k'} = \frac{1}{2}(\tau_k A_{k'} + E)$$

and the factors of the right hand side are given by

$$G_k^{(\ell)} = \left[\sqrt{\tau_k} C_{k'}^T,\ \check{A}_{k'}^T Z_k + E^T Z_k,\ \jmath \check{A}_{k'}^T Z_k,\ \jmath E^T Z_k,\ \jmath \tfrac{\sqrt{\tau_k}}{2} E^T Z_k (Z_k^T B_{k'}),\ \tfrac{\sqrt{\tau_k}}{2} E^T Z_{k+1}^{(\ell-1)} (Z_{k+1}^{(\ell-1)^T} B_{k'})\right],$$

being of column size $q + 3n_{Z_k} + 2m$.

Classical low-rank Midpoint rule for GDLEs The GALE given in a two-term factorization, associated to the application of the Midpoint rule to the GDLE (6.2), has the form

$$\hat{A}_{k'} X_{k+1} E^T + E X_{k+1} \hat{A}_{k'}^T = -\left(\tau_k B_{k'} B_{k'}^T + \tilde{A}_{k'} X_k E^T + E X_k \tilde{A}_{k'}^T\right) = -G_k G_k^T, \tag{6.21}$$

where

$$\begin{aligned}\hat{A}_{k'} &= \frac{1}{2}(\tau_k A_{k'} - E), \quad \tilde{A}_{k'} = \frac{1}{2}(\tau_k A_{k'} + E), \\ G_k &= \left[\sqrt{\tau_k} B_{k'},\ \tilde{A}_{k'} Z_k + E Z_k,\ \jmath \tilde{A}_{k'} Z_k,\ \jmath E Z_k\right] \in \mathbb{C}^{n \times (m + 3n_{Z_k})}.\end{aligned}$$

Note that for the Midpoint rule, we also have to deal with an indefinite right hand side and therefore the computations, based on the low-rank factorizations, are performed in complex arithmetic.

6.1.4. Trapezoidal Rule

Classical low-rank Trapezoidal rule for GDREs Similar to the Midpoint rule, the GARE that originates from the application of the Trapezoidal rule to the GDRE (6.1), has the form

$$\tilde{A}_{k+1}^T X_{k+1} E + E^T X_{k+1} \tilde{A}_{k+1} - E^T X_{k+1} \tilde{S}_{k+1} X_{k+1} E + \tilde{W}_{k+1} = 0 \tag{6.22}$$

with

$$\begin{aligned} \tilde{A}_{k+1} &= \frac{\tau_k}{2} A_{k+1} - \frac{1}{2} E, \quad \tilde{S}_{k+1} = \frac{\tau_k}{2} B_{k+1} B_{k+1}^T \\ \tilde{W}_{k+1} &= \frac{\tau_k}{2} C_{k+1}^T C_{k+1} + E^T X_k E + \frac{\tau_k}{2}\left(C_k^T C_k + A_k^T X_k E + E^T X_k A_k - E^T X_k S_k X_k E \right) \end{aligned}$$

and from Newton's method, we obtain the GALE

$$\begin{aligned} \hat{A}_{k+1}^{(\ell)^T} X_{k+1}^{(\ell)} E + E^T X_{k+1}^{(\ell)} \hat{A}_{k+1}^{(\ell)} = &-\frac{\tau_k}{2}(C_{k+1}^T C_{k+1} + C_k^T C_k) - \check{A}_k^T X_k E - E^T X_k \check{A}_k \\ &+ \frac{\tau_k}{2} E^T X_k B_k B_k^T X_k E - \frac{\tau_k}{2} E^T X_{k+1}^{(\ell-1)} B_{k+1} B_{k+1}^T X_{k+1}^{(\ell-1)} E = -G_k^{(\ell)} G_k^{(\ell)^T}, \end{aligned} \tag{6.23}$$

where the matrices $\hat{A}_k^{(\ell)}$, $\check{A}_k$ and the right hand side factor $G_k^{(\ell)}$ are given as

$$\begin{aligned} \hat{A}_{k+1}^{(\ell)} &= \frac{\tau_k}{2}(A_{k+1} - B_{k+1} B_{k+1}^T X_{k+1}^{(\ell-1)} E) - \frac{1}{2} E, \quad \check{A}_k = \frac{1}{2}(\tau_k A_k + E), \\ G_k^{(\ell)} &= \Big[\sqrt{\frac{\tau_k}{2}} C_{k+1}^T, \sqrt{\frac{\tau_k}{2}} C_k^T, \check{A}_k^T Z_k + E^T Z_k, \jmath \check{A}_k^T Z_k, \jmath E^T Z_k, \\ &\qquad \jmath \sqrt{\frac{\tau_k}{2}} E^T Z_k (Z_k^T B_k), \sqrt{\frac{\tau_k}{2}} E^T Z_{k+1}^{(\ell-1)} (Z_{k+1}^{(\ell-1)^T} B_{k+1}) \Big] \in \mathbb{C}^{n \times (2q + 3n_{Z_k} + 2m)}. \end{aligned}$$

Considering the autonomous GDRE with constant system matrices, we in particular have $C_{k+1} = C_k$, for all k, and the GALE (6.23) becomes

$$\begin{aligned} \hat{A}_{k+1}^{(\ell)^T} X_{k+1}^{(\ell)} E + E^T X_{k+1}^{(\ell)} \hat{A}_{k+1}^{(\ell)} &= -\left(\tau_k C^T C + \check{A}_k^T X_k E + E^T X_k \check{A}_k - \frac{\tau_k}{2} E^T X_k B B^T X_k E \right) \\ &\quad - \frac{\tau_k}{2} E^T X_{k+1}^{(\ell-1)} B B^T X_{k+1}^{(\ell-1)} E = -G_k^{(\ell)} G_k^{(\ell)^T} \end{aligned}$$

with

$$\begin{aligned} \hat{A}_{k+1}^{(\ell)} &= \frac{\tau_k}{2}(A - B B^T X_{k+1}^{(\ell-1)} E) - \frac{1}{2} E, \quad \check{A}_k = \frac{1}{2}(\tau_k A + E) \\ G_k^{(\ell)} &= \Big[\sqrt{\tau_k} C^T, \check{A}_k^T Z_k + E^T Z_k, \jmath \check{A}_k^T Z_k, \jmath E^T Z_k, \\ &\qquad \jmath \sqrt{\frac{\tau_k}{2}} E^T Z_k (Z_k^T B), \sqrt{\frac{\tau_k}{2}} E^T Z_{k+1}^{(\ell-1)} (Z_{k+1}^{(\ell-1)^T} B) \Big] \in \mathbb{C}^{n \times (q + 3n_{Z_k} + 2m)}. \end{aligned}$$

That is, n_{G_k} is shortened by q columns, compared to the non-autonomous case.

Classical low-rank Trapezoidal rule for GDLEs In case of the GDLE (6.2), the Trapezoidal rule yields the GALE

$$\begin{aligned} \hat{A}_{k+1}^{(\ell)^T} X_{k+1}^{(\ell)} E + E^T X_{k+1}^{(\ell)} \hat{A}_{k+1}^{(\ell)} &= -\left(\frac{\tau_k}{2}(B_{k+1}B_{k+1}^T + B_k B_k^T) + \tilde{A}_k X_k E^T + E X_k \tilde{A}_k^T\right) \\ &= -G_k^{(\ell)} G_k^{(\ell)^T} \end{aligned}$$

with

$$\begin{aligned} \hat{A}_{k+1}^{(\ell)} &= \frac{1}{2}(\tau_k A_{k+1} - E), \quad \tilde{A}_k = \frac{1}{2}(\tau_k A_k + E) \\ G_k^{(\ell)} &= \left[\sqrt{\tfrac{\tau_k}{2}} B_{k+1}, \sqrt{\tfrac{\tau_k}{2}} B_k, \tilde{A}_k Z_k + E Z_k, \jmath \tilde{A}_k Z_k, \jmath E Z_k\right] \in \mathbb{C}^{n \times (2m+3n_{Z_k})}. \end{aligned}$$

Then, for the autonomous GDLE with $B_{k+1} = B_k$, we obtain the GALE

$$\hat{A}_{k+1}^{(\ell)^T} X_{k+1}^{(\ell)} E + E^T X_{k+1}^{(\ell)} \hat{A}_{k+1}^{(\ell)} = -\left(\tau_k B B^T + \tilde{A}_k X_k E^T + E X_k \tilde{A}_k^T\right) = -G_k^{(\ell)} G_k^{(\ell)^T}$$

and, consequently, the right hand side factor reads

$$\begin{aligned} \hat{A}_{k+1}^{(\ell)} &= \frac{1}{2}(\tau_k A - E), \quad \tilde{A}_k = \frac{1}{2}(\tau_k A + E), \\ G_k^{(\ell)} &= \left[\sqrt{\tau_k} B, \tilde{A}_k Z_k + E Z_k, \jmath \tilde{A}_k Z_k, \jmath E Z_k\right] \in \mathbb{C}^{n \times (m+3n_{Z_k})}. \end{aligned}$$

6.1.5. Peer Methods

Classical low-rank implicit Peer scheme for GDREs The implicit peer scheme (5.25) applied to the GDRE (6.1) results in the solution of the GARE

$$\tilde{A}_{k,i}^T X_{k,i} E + E^T X_{k,i} \tilde{A}_{k,i} - X_{k,i} \tilde{S}_{k,i} X_{k,i} + \tilde{W}_{k,i} = 0, \qquad i = 1, \ldots, s$$

with

$$\begin{aligned} \tilde{A}_{k,i} &= \tau_k g_{i,i} A_{k,i} - \frac{1}{2} E, \quad \tilde{S}_{k,i} = \tau_k g_{i,i} B_{k,i} B_{k,i}^T, \\ \tilde{W}_{k,i} &= \tau_k g_{i,i} C_{k,i}^T C_{k,i} + \sum_{j=1}^{s} b_{i,j} E^T X_{k-1,j} E + \tau_k \sum_{j=1}^{s} a_{i,j} \mathcal{R}(t_{k-1,j}, X_{k-1,j}) + \tau_k \sum_{j=1}^{i-1} g_{i,j} \mathcal{R}(t_{k,j}, X_{k,j}) \end{aligned}$$

and the application of Newton's method yields the GALE

$$\hat{A}_{k,i}^{(\ell)^T} X_{k,i}^{(\ell)} E + E^T X_{k,i}^{(\ell)} \hat{A}_{k,i}^{(\ell)} = -\tilde{W}_{k,i} - \tau_k g_{i,i} E^T X_{k,i}^{(\ell-1)} B_{k,i} B_{k,i}^T X_{k,i}^{(\ell-1)} E = -G_{k,i}^{(\ell)} G_{k,i}^{(\ell)^T} \tag{6.24}$$

with

$$\hat{A}_{k,i}^{(\ell)} = \tau_k g_{i,i} (A_{k,i} - B_{k,i} B_{k,i}^T X_{k,i}^{(\ell-1)} E) - \frac{1}{2} E.$$

Now, assuming $X_{k,j}$ admits a decomposition $X_{k,j} = Z_{k,j}Z_{k,j}^T$ with $Z_{k,j} \in \mathbb{C}^{n\times n_{Z_{k,j}}}$, $\forall k, j$ and similarly applying the factorization used for (6.12) in the Rosenbrock methods, to the Riccati operator $\mathcal{R}(.,.)$, we obtain

$$\begin{aligned}\mathcal{R}(t_{k,j}, X_{k,j}) &= C_{k,j}^T C_{k,j} + A_{k,j}^T X_{k,j}E + E^T X_{k,j}A_{k,j} - E^T X_{k,j}B_{k,j}B_{k,j}^T X_{k,j}E = \mathcal{Z}_{k,j}\mathcal{Z}_{k,j}^T,\\ \mathcal{Z}_{k,j} &= \left[C_{k,j}^T,\ A_{k,j}^T Z_{k,j} + E^T Z_{k,j},\ \jmath A_{k,j}^T Z_{k,j},\ \jmath E^T Z_{k,j},\ \jmath E^T X_{k,j}B_{k,j}\right]\end{aligned} \tag{6.25}$$

with $\mathcal{Z}_{k,j} \in \mathbb{C}^{n\times(q+3n_{Z_{k,j}}+m)}$. Finally, the right hand side factor $G_{k,i}^{(\ell)}$ can be written in the form

$$\begin{aligned}G_{k,i}^{(\ell)} = \Big[&\sqrt{\tau_k g_{i,i}}C_{k,i}^T,\ \sqrt{b_{i,1}}E^T Z_{k-1,1}, \dots,\ \sqrt{b_{i,s}}E^T Z_{k-1,s},\\ &\sqrt{\tau_k a_{i,1}}\mathcal{Z}_{k-1,1}, \dots,\ \sqrt{\tau_k a_{i,s}}\mathcal{Z}_{k-1,s},\\ &\sqrt{\tau_k g_{i,1}}\mathcal{Z}_{k,1}, \dots,\ \sqrt{\tau_k g_{i,i-1}}\mathcal{Z}_{k,i-1},\ \sqrt{\tau_k g_{i,i}}E^T Z_{k,i}^{(\ell-1)}(Z_{k,i}^{(\ell-1)^T} B_{k,i})\Big]\end{aligned} \tag{6.26}$$

and thus is of column size

$$\begin{aligned}n_{G_{k,i}} &= q + \sum_{j=1}^{s} n_{Z_{k-1,j}} + \sum_{j=1}^{s}(q + 3n_{Z_{k-1,j}} + m) + \sum_{j=1}^{i-1}(q + 3n_{Z_{k,j}} + m) + m\\ &= (s+i)(q+m) + 4\sum_{j=1}^{s} n_{Z_{k-1,j}} + 3\sum_{j=1}^{i-1} n_{Z_{k,j}},\ i = 1, \dots, s.\end{aligned} \tag{6.27}$$

For time invariant and therefore autonomous systems with constant system matrices, in particular the $C_{k,i}$ are equal for all $k,\ i$. Hence, the GALE to be solved reads

$$\hat{A}_{k,i}^{(\ell)^T} X_{k,i}^{(\ell)}E + E^T X_{k,i}^{(\ell)}\hat{A}_{k,i}^{(\ell)} = -\tilde{W}_{k,i} - \tau_k g_{i,i}E^T X_{k,i}^{(\ell-1)}BB^T X_{k,i}^{(\ell-1)}E = -G_{k,i}^{(\ell)}G_{k,i}^{(\ell)^T}, \tag{6.28}$$

where $\hat{A}_{k,i}$ and the right hand side factor $G_{k,i}^{(\ell)}$ are given by

$$\begin{aligned}\hat{A}_{k,i}^{(\ell)} &= \tau_k g_{i,i}(A - BB^T X_{k,i}^{(\ell-1)}E) - \frac{1}{2}E\\ G_{k,i}^{(\ell)} &= \Big[\sqrt{\sum_{j=1}^{s}\tau_k a_{i,j} + \sum_{j=1}^{i}\tau_k g_{i,j}}\ C^T,\ \sqrt{b_{i,1}}E^T Z_{k-1,1}, \dots,\ \sqrt{b_{i,s}}E^T Z_{k-1,s},\\ &\quad \sqrt{\tau_k a_{i,1}}\mathcal{Z}_{k-1,1}, \dots,\ \sqrt{\tau_k a_{i,s}}\mathcal{Z}_{k-1,s},\\ &\quad \sqrt{\tau_k g_{i,1}}\mathcal{Z}_{k,1}, \dots,\ \sqrt{\tau_k g_{i,i-1}}\mathcal{Z}_{k,i-1},\ \sqrt{\tau_k g_{i,i}}E^T Z_{k,i}^{(\ell-1)}(Z_{k,i}^{(\ell-1)^T} B)\Big],\end{aligned}$$

while the factors $\mathcal{Z}_{k,j}$ become

$$\mathcal{Z}_{k,j} = \left[A^T Z_{k,j} + E^T Z_{k,j},\ \jmath A^T Z_{k,j},\ \jmath E^T Z_{k,j},\ \jmath E^T X_{k,j}B\right] \in \mathbb{C}^{n\times(3n_{Z_{k,j}}+m)}.$$

That is, the column size of $G_{k,i}^{(\ell)}$ reduces by $(s+i-1)q$ and is given by

$$q + (s+i)m + 4\sum_{j=1}^{s} n_{Z_{k-1,j}} + 3\sum_{j=1}^{i-1} n_{Z_{k,j}},\ i = 1,\dots,s.$$

With respect to all peer stages, this is a reduction of $s^2 + \sum_{i=1}^{s} i - s = \frac{3s^2-s}{2}$ right hand side columns, entering the Lyapunov solver at each time step of an s-stage implicit peer integration method (5.25), in total.

Classical low-rank implicit Peer scheme for GDLEs For the GDLE (6.2), the GALE to be solved is of the form

$$\tilde{A}_{k,i}X_{k,i}E^T + EX_{k,i}\tilde{A}_{k,i}^T = -\tilde{W}_{k,i} = -G_{k,i}G_{k,i}^T \tag{6.29}$$

with

$$\begin{aligned}\tilde{A}_{k,i}^{(\ell)} &= \tau_k g_{i,i}A_{k,i} - \frac{1}{2}E,\\ \tilde{W}_{k,i} &= \tau_k g_{i,i}B_{k,i}B_{k,i}^T + \sum_{j=1}^{s} b_{i,j}EX_{k-1,j}E^T + \tau_k\sum_{j=1}^{s} a_{i,j}\mathcal{L}(t_{k-1,j}, X_{k-1,j}) + \tau_k\sum_{j=1}^{i-1} g_{i,j}\mathcal{L}(t_{k,j}, X_{k,j}).\end{aligned}$$

In analogy to (6.25), we obtain the low-rank representation

$$\begin{aligned}\mathcal{L}(t_{k,j}, X_{k,j}) &= B_{k,j}B_{k,j}^T + A_{k,j}X_{k,j}E^T + EX_{k,j}A_{k,j}^T = \mathcal{Z}_{k,j}\mathcal{Z}_{k,j}^T,\\ \mathcal{Z}_{k,j} &= \left[B_{k,j},\ A_{k,j}Z_{k,j} + EZ_{k,j},\ \jmath A_{k,j}Z_{k,j},\ \jmath EZ_{k,j}\right] \in \mathbb{C}^{n\times(m+3n_{Z_{k,j}})}\end{aligned}$$

for the Lyapunov operator $\mathcal{L}$. Finally, the right hand side factor reads

$$\begin{aligned}G_{k,i} = \Big[&\sqrt{\tau_k g_{i,i}}B_{k,i},\ \sqrt{b_{i,1}}E^TZ_{k-1,1},\dots,\ \sqrt{b_{i,s}}E^TZ_{k-1,s}\\ &\sqrt{\tau_k a_{i,1}}\mathcal{Z}_{k-1,1},\dots,\ \sqrt{\tau_k a_{i,s}}\mathcal{Z}_{k-1,s},\ \sqrt{\tau_k g_{i,1}}\mathcal{Z}_{k,1},\dots,\ \sqrt{\tau_k g_{i,i-1}}\mathcal{Z}_{k,i-1}\Big]\end{aligned} \tag{6.30}$$

and is of column size

$$(s+i)m + 4\sum_{j=1}^{s} n_{Z_{k-1,j}} + 3\sum_{j=1}^{i-1} n_{Z_{k,j}},\ i = 1,\dots,s.$$

In the autonomous case, the GALE (6.29) becomes

$$\tilde{A}_{k,i}X_{k,i}E^T + EX_{k,i}\tilde{A}_{k,i}^T = -\tilde{W}_{k,i} = -G_{k,i}G_{k,i}^T \tag{6.31}$$

with

$$\begin{aligned}\tilde{A}_{k,i} &= \tau_k g_{i,i}A - \frac{1}{2}E,\\ \tilde{W}_{k,i} &= \tau_k g_{i,i}BB^T + \sum_{j=1}^{s} b_{i,j}EX_{k-1,j}E^T + \tau_k\sum_{j=1}^{s} a_{i,j}\mathcal{L}(X_{k-1,j}) + \tau_k\sum_{j=1}^{i-1} g_{i,j}\mathcal{L}(X_{k,j}),\end{aligned}$$

and for the factors of the right hand side, we write

$$G_{k,i} = \Big[\sqrt{\sum_{j=1}^{s} \tau_k a_{i,j} + \sum_{j=1}^{i} \tau_k g_{i,j}}\, B,\ \sqrt{b_{i,1}} E^T Z_{k-1,1}, \ldots, \sqrt{b_{i,s}} E^T Z_{k-1,s},$$
$$\sqrt{\tau_k a_{i,1}} \mathcal{Z}_{k-1,1}, \ldots, \sqrt{\tau_k a_{i,s}} \mathcal{Z}_{k-1,s}, \sqrt{\tau_k g_{i,1}} \mathcal{Z}_{k,1}, \ldots, \sqrt{\tau_k g_{i,i-1}} \mathcal{Z}_{k,i-1} \Big],$$

where

$$\mathcal{Z}_{k,j} = \left[AZ_{k,j} + EZ_{k,j},\ \jmath AZ_{k,j},\ \jmath EZ_{k,j} \right]$$

and $G_{k,i}$ is of column size

$$m + 4 \sum_{j=1}^{s} n_{Z_{k-1,j}} + 3 \sum_{j=1}^{i-1} n_{Z_{k,j}}.$$

That is, the number of columns in $G_{k,i}$ reduces by $(s + i - 1)m$.

Classical low-rank Rosenbrock-type Peer scheme for GDREs The application of the linear implicit Rosenbrock-type peer scheme (5.30) to the GDRE (6.1) results in the solution of the GALE

$$\tilde{A}_{k,i}^T X_{k,i} E + E^T X_{k,i} \tilde{A}_{k,i} = -\tilde{W}_{k,i}, \quad i = 1, \ldots, s, \tag{6.32}$$

with

$$\tilde{W}_{k,i} = \sum_{j=1}^{s} b_{i,j} E^T X_{k-1,j} E + \tau_k \sum_{j=1}^{s} a_{i,j} \left(\mathcal{R}(t_{k-1,j}, X_{k-1,j}) - (\hat{A}_k^T X_{k-1,j} E + E^T X_{k-1,j} \hat{A}_k) \right),$$
$$+ \tau_k \sum_{j=1}^{i-1} g_{i,j} (\hat{A}_k^T X_{k,j} E + E^T X_{k,j} \hat{A}_k),$$
$$\hat{A}_k = A_k - B_k B_k^T X_k E, \quad \tilde{A}_{k,i} = \tau_k g_{i,i} \hat{A}_k - \frac{1}{2} E.$$

In contrast to small-scale and dense computations it is recommended to never explicitly form the matrices $\hat{A}_k$. Therefore, instead we use

$$\hat{A}_k^T X_{k-1,j} E + E^T X_{k-1,j} \hat{A}_k = A_k^T X_{k-1,j} E + E^T X_{k-1,j} A_k$$
$$- E^T X_k B_k B_k^T X_{k-1,j} E - E^T X_{k-1,j} B_k B_k^T X_k E.$$

Moreover, exploiting the structure of the Riccati operators $\mathcal{R}(t_{k-1,j}, X_{k-1,j})$, the right hand side $\tilde{W}_{k,i}$ can be reformulated in the form

$$\begin{aligned}\tilde{W}_{k,i} = \tau_k \sum_{j=1}^{i-1} g_{i,j}\left(A_k^T X_{k,j} E + E^T X_{k,j} A_k - E^T X_k B_k B_k^T X_{k,j} E - E^T X_{k,j} B_k B_k^T X_k E\right) \\ + \sum_{j=1}^{s}\Big(\tau_k a_{i,j}\big(C_{k-1,j}^T C_{k-1,j} - E^T X_{k-1,j} B_{k-1,j} B_{k-1,j}^T X_{k-1,j} E \\ + E^T X_k B_k B_k^T X_{k-1,j} E + E^T X_{k-1,j} B_k B_k^T X_k E\big) + \check{A}_{k,i,j}^T X_{k-1,j} E + E^T X_{k-1,j} \check{A}_{k,i,j}\Big),\end{aligned}$$

where $\check{A}_{k,i,j} = \tau_k a_{i,j}(A_{k-1,j} - A_k) + \frac{b_{i,j}}{2} E$. The matrix $\check{A}_{k,i,j}$ can efficiently be computed, since $A_{k-1,j}$, A_k and E are sparse matrices and so is $\check{A}_{k,i,j}$. Note that for $j = s$, we have $A_{k-1,s} = A_k$, $B_{k-1,s} = B_k$ and $X_{k-1,s} = X_k$. Therefore $\check{A}_{k,i,s} = \frac{b_{i,s}}{2} E$ and the right hand side at every stage $i = 1, \ldots, s$ reduces to

$$\begin{aligned}\tilde{W}_{k,i} = \tau_k \sum_{j=1}^{i-1} g_{i,j}\left(A_k^T X_{k,j} E + E^T X_{k,j} A_k - E^T X_k B_k B_k^T X_{k,j} E - E^T X_{k,j} B_k B_k^T X_k E\right) \\ + \sum_{j=1}^{s-1}\Big(\tau_k a_{i,j}\big(C_{k-1,j}^T C_{k-1,j} - E^T X_{k-1,j} B_{k-1,j} B_{k-1,j}^T X_{k-1,j} E \\ + E^T X_k B_k B_k^T X_{k-1,j} E + E^T X_{k-1,j} B_k B_k^T X_k E\big) + \check{A}_{k,i,j}^T X_{k-1,j} E + E^T X_{k-1,j} \check{A}_{k,i,j}\Big) \\ + \tau_k a_{i,s}\left(C_k^T C_k + E^T X_k B_k B_k^T X_k E\right) + b_{i,s} E^T X_k E.\end{aligned}$$

Also, we see that a considerable number of quadratic terms share the product $E^T X_k B_k$ or its transposed. Combining these expressions, we obtain the formulation

$$\begin{aligned}\tilde{W}_{k,i} = \tau_k \sum_{j=1}^{i-1} g_{i,j}\left(A_k^T X_{k,j} E + E^T X_{k,j} A_k\right) + E^T X_k B_k K_{k,i}^T + K_{k,i} B_k^T X_k E \\ + \sum_{j=1}^{s-1}\Big(\tau_k a_{i,j}\big(C_{k-1,j}^T C_{k-1,j} - E^T X_{k-1,j} B_{k-1,j} B_{k-1,j}^T X_{k-1,j} E\big) \\ + \check{A}_{k,i,j}^T X_{k-1,j} E + E^T X_{k-1,j} \check{A}_{k,i,j}\Big) + \tau_k a_{i,s} C_k^T C_k + b_{i,s} E^T X_k E,\end{aligned} \tag{6.33}$$

where

$$K_{k,i} = \tau_k E^T \left(\sum_{j=1}^{s-1} a_{i,j} X_{k-1,j} + \frac{a_{i,s}}{2} X_k - \sum_{j=1}^{i-1} g_{i,j} X_{k,j}\right) B_k$$

collects all products, interacting with $E^T X_k B_k$. Here, $X_{k-1,j}$, X_k and $X_{k,j}$ are assumed to be given in low-rank format with the factors $Z_{k-1,j}$, $Z_k = Z_{k-1,s}$, $Z_{k,j}$, $j = 1,\ldots,s$, respectively. Thus, forming the single summands and therefore $K_{k,i} \in \mathbb{R}^{n\times m}$, $m \ll n$, is computationally cheap compared to the column compression technique that should be applied to the factor, resulting from the separated summands. Finally, defining the factors

$$\begin{aligned}
\mathcal{Z}_{k,j} &= \left[A_k^T Z_{k,j} + E^T Z_{k,j},\ \jmath A_k^T Z_{k,j},\ \jmath E^T Z_{k,j}\right] \in \mathbb{C}^{n\times 3n_{Z_{k,j}}},\\
\hat{\mathcal{Z}}_{k-1,j} &= \left[C_{k-1,j}^T,\ \jmath E^T X_{k-1,j} B_{k-1,j}\right] \in \mathbb{C}^{n\times(q+m)},\\
\check{\mathcal{Z}}_{k,i,j} &= \left[\ \check{A}_{k,i,j}^T Z_{k-1,j} + E^T Z_{k-1,j},\ \jmath \check{A}_{k,i,j}^T Z_{k-1,j},\ \jmath E^T Z_{k-1,j}\right] \in \mathbb{C}^{n\times(3n_{Z_{k-1,j}})},
\end{aligned}$$

the classical low-rank factorization $\tilde{W}_{k,i} = G_{k,i} G_{k,i}^T$ of (6.33) is given by the factor

$$\begin{aligned}
G_{k,i} = \Big[&\sqrt{\tau_k g_{i,1}}\mathcal{Z}_{k,1},\ldots,\ \sqrt{\tau_k g_{i,i-1}}\mathcal{Z}_{k,i-1},\ E^T X_k B_k + K_{k,i},\ \jmath E^T X_k B_k,\ \jmath K_{k,i}\\
&\sqrt{\tau_k a_{i,1}}\hat{\mathcal{Z}}_{k-1,1},\ldots,\ \sqrt{\tau_k a_{i,s-1}}\hat{\mathcal{Z}}_{k-1,s-1},\\
&\check{\mathcal{Z}}_{k,i,1},\ldots,\check{\mathcal{Z}}_{k,i,s-1},\ \sqrt{\tau_k a_{i,s}} C_k^T,\ \sqrt{b_{i,s}} E^T Z_{k-1,s}\Big],\ i = 1,\ldots,s
\end{aligned} \tag{6.34}$$

and is of column size

$$\begin{aligned}
&\sum_{j=1}^{i-1} 3n_{Z_{k,j}} + 3m + \sum_{j=1}^{s-1}(q+m) + \sum_{j=1}^{s-1} 3n_{Z_{k-1,j}} + q + n_{Z_{k-1,s}}\\
&= 3\sum_{j=1}^{i-1} n_{Z_{k,j}} + 3\sum_{j=1}^{s-1} n_{Z_{k-1,j}} + n_{Z_{k-1,s}} + sq + (s+2)m.
\end{aligned} \tag{6.35}$$

In the autonomous case, we in particular have $A_{k-1,j} = A_k = A$. Hence, $\check{A}_{k,i,j} = \frac{b_{i,j}}{2}E$, $i,j = 1,\ldots,s$, and together with the modifications for $j = s$, $X_{k-1,s} = X_k$, the right hand side $\tilde{W}_{k,i}$ in (6.33) becomes

$$\begin{aligned}
\tilde{W}_{k,i} =&\tau_k \sum_{j=1}^{i-1} g_{i,j}\left(A_k^T X_{k,j} E + E^T X_{k,j} A_k\right) + E^T X_k B_k K_{k,i}^T + K_{k,i} B_k^T X_k E\\
&+ \sum_{j=1}^{s}\left(\tau_k a_{i,j} C^T C + b_{i,j} E^T X_{k-1,j} E\right) - \sum_{j=1}^{s-1} \tau_k a_{i,j} E^T X_{k-1,j} B B^T X_{k-1,j} E.
\end{aligned} \tag{6.36}$$

Then, similar to the non-autonomous factorization, we obtain the right hand side factor

$$\begin{aligned}
G_{k,i} = \Big[&\ \sqrt{\tau_k g_{i,1}}\mathcal{Z}_{k,1},\ldots,\ \sqrt{\tau_k g_{i,i-1}}\mathcal{Z}_{k,i-1},\ E^T X_k B + K_{k,i},\ \jmath E^T X_k B,\ \jmath K_{k,i}\\
&\sqrt{\tau_k \sum_{j=1}^{s} a_{i,j}} C^T,\ \sqrt{b_{i,1}} E^T Z_{k-1,1},\ldots,\ \sqrt{b_{i,s}} E^T Z_{k-1,s},\\
&\jmath\sqrt{\tau_k a_{i,1}} E^T X_{k-1,1} B,\ldots,\jmath\sqrt{\tau_k a_{i,s-1}} E^T X_{k-1,s-1} B\Big],\ i = 1,\ldots,s
\end{aligned} \tag{6.37}$$

with

$$\mathcal{Z}_{k,j} = \left[A^T Z_{k,j} + E^T Z_{k,j},\ \jmath A^T Z_{k,j},\ \jmath E^T Z_{k,j}\right] \in \mathbb{C}^{n \times 3n_{Z_{k,j}}}.$$

That is, $G_{k,i}, i = 1, \ldots, s$, is of column size

$$\begin{aligned}
&\sum_{j=1}^{i-1} 3n_{Z_{k,j}} + 3m + q + \sum_{j=1}^{s} n_{Z_{k-1,j}} + \sum_{j=1}^{s-1} m \\
&= 3\sum_{j=1}^{i-1} n_{Z_{k,j}} + \sum_{j=1}^{s} n_{Z_{k-1,j}} + q + (s+2)m.
\end{aligned} \tag{6.38}$$

Now, considering the modified Rosenbrock-type peer formulation (5.37), for the GDRE (6.1), we obtain the GALE

$$\begin{aligned}
\tilde{A}_{k,i}^T Y_{k,i} E + E^T Y_{k,i} \tilde{A}_{k,i} &= -\tilde{W}_{k,i}, \quad i = 1, \ldots, s, \\
\tilde{W}_{k,i} &= \sum_{j=1}^{s} \frac{\mathbf{b}_{i,j}}{\tau_k} E^T Y_{k-1,j} E + \sum_{j=1}^{s} a_{i,j} \mathcal{R}(t_{k-1,j}, \sum_{\ell=1}^{j} \mathbf{g}_{j,\ell} Y_{k-1,\ell}) \\
&\quad - \sum_{j=1}^{s} \mathbf{a}_{i,j} (\hat{A}_k^T Y_{k-1,j} E + E^T Y_{k-1,j} \hat{A}_k) - \sum_{j=1}^{i-1} \frac{\mathbf{g}_{i,j}}{\tau_k} E^T Y_{k,j} E, \\
\tilde{A}_{k,i} &= \hat{A}_k - \frac{1}{2\tau_k g_{i,i}} E, \quad \hat{A}_k = A_k - B_k B_k^T X_k E.
\end{aligned} \tag{6.39}$$

Note that the matrix $\hat{A}_k$ is given in terms of X_k. This is due to the fact that $\hat{A}_k$ originates from the Jacobian (5.10) that, as in the original scheme, is given as the Fréchet derivative of $\mathcal{R}(t_k, X_k) = \mathcal{R}(t_{k-1,s}, X_{k-1,s}) = \mathcal{R}(t_{k-1,s}, \sum_{\ell=1}^{s} \mathbf{g}_{j,\ell} Y_{k-1,\ell})$. As previously mentioned in Section 5.4.2, the solution approximations $X_{k,j}$ have to be reconstructed from the auxiliary variables $Y_{k,j}$. Therefore, the original variables $X_{k-1,j}$ (with $X_k = X_{k-1,s}$) within $\hat{A}_k$, as well as in the Riccati operators $\mathcal{R}(t_{k-1,j}, X_{k-1,j})$ are kept throughout the computations. That is, using both sets of variables does not require additional computations. In order to give a more detailed motivation for mixing up the original and auxiliary scheme, the following considerations are stated.

From the relation of the original and auxiliary variables, given in (5.33), we have $X_{k-1,j} = \sum_{\ell=1}^{j} \mathbf{g}_{j,\ell} Y_{k-1,\ell}$. Further, defining the decomposition $Y_{k-1,\ell} = \hat{Z}_{k-1,\ell} \hat{Z}_{k-1,\ell}^T$, $\hat{Z}_{k-1,\ell} \in \mathbb{R}^{n \times n_{\hat{Z}_{k-1,\ell}}}$, $\ell = 1, \ldots, j$, the original solution approximation admits a factorization $X_{k-1,j} = Z_{k-1,j} Z_{k-1,j}^T$, $j = 1, \ldots, s$, based on the factors

$$Z_{k-1,j} = \left[\sqrt{\mathbf{g}_{j,1}} \hat{Z}_{k-1,1}, \ldots, \sqrt{\mathbf{g}_{j,j}} \hat{Z}_{k-1,j}\right] \in \mathbb{C}^{n \times n_{Z_{k-1,j}}}.$$

The factors $Z_{k-1,j}$ are given as a block concatenation of the solution factors of the auxiliary variables $Y_{k-1,\ell}$, $\ell = 1, \ldots, j$. Thus, the column size $n_{Z_{k-1,j}} = \sum_{\ell=1}^{j} n_{\hat{Z}_{k-1,\ell}}$ may dramatically grow with respect to the number of stages and time steps. Still, the numerical rank of the original solution is assumed to be "small". That is, using the column compression techniques, being a tacit requirement for large-scale problems anyway, the column size of $Z_{k-1,j}$ is presumably "small" as well. To be more precise, the factors $Z_{k-1,j}$ and $\hat{Z}_{k-1,j}$ are expected to be of comparable size. As a consequence, one can make use of both representations at the one place or another without messing up the formulations with respect to both, the notational and computational complexity.

However, expanding $\hat{A}_k$ and combining the linear parts with respect to $Y_{k-1,j}$, the right hand side reads

$$\begin{aligned}\tilde{W}_{k,i} = &-\sum_{j=1}^{s}\Big(\check{A}_{k,i,j}^T Y_{k-1,j}E + E^T Y_{k-1,j}\check{A}_{k,i,j} - \mathbf{a}_{i,j}\left(E^T X_k B_k B_k^T Y_{k-1,j}E + E^T Y_{k-1,j} B_k B_k^T X_k E\right)\Big)\\ &+ \sum_{j=1}^{s} a_{i,j}\mathcal{R}(t_{k-1,j}, X_{k-1,j}) - \sum_{j=1}^{i-1}\frac{\mathbf{g}_{i,j}}{\tau_k}E^T Y_{k,j}E\end{aligned} \tag{6.40}$$

with $\check{A}_{k,i,j} = \mathbf{a}_{i,j}A_k - \frac{\mathbf{b}_{i,j}}{2\tau_k}E$. Then, separating $\mathcal{R}(t_{k-1,s}, X_{k-1,s}) = \mathcal{R}(t_k, X_k)$ and again combining the quadratic terms containing the products $E^T X_k B_k$, we end up with the formulation

$$\begin{aligned}\tilde{W}_{k,i} = &-\sum_{j=1}^{s}\Big(\check{A}_{k,i,j}^T Y_{k-1,j}E + E^T Y_{k-1,j}\check{A}_{k,i,j}\Big) + E^T X_k B_k K_{k,i}^T + K_{k,i}B_k^T X_k E\\ &+ \sum_{j=1}^{s-1} a_{i,j}\mathcal{R}(t_{k-1,j}, X_{k-1,j}) + a_{i,s}\left(C_k^T C_k + A_k^T X_k E + E^T X_k A_k\right)\\ &- \sum_{j=1}^{i-1}\frac{\mathbf{g}_{i,j}}{\tau_k}E^T Y_{k,j}E,\\ K_{k,i} = &\, E^T\left(\sum_{j=1}^{s}\mathbf{a}_{i,j}Y_{k-1,j} - \frac{a_{i,s}}{2}X_k\right)B_k.\end{aligned} \tag{6.41}$$

In analogy to (6.13), we define the factors

$$\check{\mathcal{Z}}_{k-1,i,j} = \left[\check{A}_{k,i,j}^T\hat{Z}_{k-1,j} + E^T\hat{Z}_{k-1,j},\ \jmath\check{A}_{k,i,j}^T\hat{Z}_{k-1,j},\ \jmath E^T\hat{Z}_{k-1,j}\right] \in \mathbb{C}^{n\times 3n_{\hat{Z}_{k-1,j}}},$$

representing the low-rank factorization of

$$\check{A}_{k,i,j}^T Y_{k-1,j}E + E^T Y_{k-1,j}\check{A}_{k,i,j}$$

and following the decomposition of (6.12), one obtains the expressions

$$\begin{aligned}\mathcal{R}(t_{k-1,j}, X_{k-1,j}) &= \mathcal{Z}_{k-1,j}\mathcal{Z}_{k-1,j}^T,\\ \mathcal{Z}_{k-1,j} &= \left[C_{k-1,j}^T,\ A_{k-1,j}^T Z_{k-1,j} + E^T Z_{k-1,j},\ \jmath A_{k-1,j}^T Z_{k-1,j},\ \jmath E^T Z_{k-1,j},\ \jmath E^T Z_{k-1,j}(Z_{k-1,j}^T B_{k-1,j})\right]\end{aligned}$$

for the Riccati operators of the modified scheme with $\mathcal{Z}_{k-1,j} \in \mathbb{C}^{n\times(q+3n_{Z_{k-1,j}}+m)}$. Finally, the searched for right hand side factor is given by

$$
\begin{aligned}
G_{k,i} = \Big[& \jmath\check{\mathcal{Z}}_{k-1,i,1}, \ldots, \jmath\check{\mathcal{Z}}_{k-1,i,s}, E^T X_k B_k + K_{k,i}, \jmath E^T X_k B_k, \jmath K_{k,i} \\
& \sqrt{a_{i,1}}\mathcal{Z}_{k-1,1}, \ldots, \sqrt{a_{i,s-1}}\mathcal{Z}_{k-1,s-1}, \sqrt{a_{i,s}}C_k^T, \sqrt{a_{i,s}}(A_k^T Z_k + E^T Z_k), \jmath\sqrt{a_{i,s}}A_k^T Z_k, \jmath\sqrt{a_{i,s}}E^T Z_k, \\
& \jmath\sqrt{\frac{\mathbf{g}_{i,1}}{\tau_k}}E^T\hat{Z}_{k,1}, \ldots, \jmath\sqrt{\frac{\mathbf{g}_{i,i-1}}{\tau_k}}E^T\hat{Z}_{k,i-1}, \Big],
\end{aligned}
$$

and is of column size

$$
\begin{aligned}
& \sum_{j=1}^{s} 3n_{\hat{Z}_{k-1,j}} + 3m + \sum_{j=1}^{s-1}(q + 3n_{Z_{k-1,j}} + m) + q + 3n_{Z_{k-1,s}} + \sum_{j=1}^{i-1} n_{\hat{Z}_{k,j}} \\
&= \sum_{j=1}^{i-1} n_{\hat{Z}_{k,j}} + 3\sum_{j=1}^{s}\left(n_{\hat{Z}_{k-1,j}} + n_{Z_{k-1,j}}\right) + sq + (s+2)m.
\end{aligned}
\tag{6.42}
$$

Note that the use of both, the auxiliary variables in the linear parts and the original variables within the Fréchet derivative and the Riccati operator, does not allow us to completely combine these parts, as we have seen for the condensed form (6.33) of the original Rosenbrock-type peer scheme. Therefore, assume the associated low-rank factors $Z_{k-1,j}$ and $\hat{Z}_{k-1,j}$ of $X_{k-1,j}$ and $Y_{k-1,j}$, respectively, to be of comparable column sizes $n_{Z_{k-1,j}}$ and $n_{\hat{Z}_{k-1,j}}$. Then, comparing (6.35) and (6.42), the modified scheme results in a larger number of columns in the right hand side factorization, although avoiding the application of the Jacobian to the current solutions $Y_{k,j}$, $j = 1, \ldots, i$, saves $2\sum_{j=1}^{i-1} n_{\hat{Z}_{k,j}}$ columns. That is, for large-scale non-autonomous (G)DREs, the standard version of the Rosenbrock-type peer schemes seems to be preferable. An insightful comparison based on numerical experiments is given in Section 6.4.

Still, a more beneficial situation can be found for autonomous GDREs. Here, additional modifications, based on the time-invariant nature of the system matrices, allow to further reduce the complexity of the GALEs to be solved. In that case the associated GALEs are of the form

$$
\begin{aligned}
& \tilde{A}_{k,i}^T Y_{k,i} E + E^T Y_{k,i} \tilde{A}_{k,i} = -\tilde{W}_{k,i} \quad i = 1, \ldots, s, \\
& \tilde{W}_{k,i} = \sum_{j=1}^{s} \frac{\mathbf{b}_{i,j}}{\tau_k} E^T Y_{k-1,j} E + \sum_{j=1}^{s} a_{i,j} \mathcal{R}(\sum_{\ell=1}^{j} \mathbf{g}_{j,\ell} Y_{k-1,\ell}) \\
& \qquad - \sum_{j=1}^{s} \mathbf{a}_{i,j}(\hat{A}_k^T Y_{k-1,j} E + E^T Y_{k-1,j} \hat{A}_k) - \sum_{j=1}^{i-1} \frac{\mathbf{g}_{i,j}}{\tau_k} E^T Y_{k,j} E, \\
& \tilde{A}_{k,i} = \hat{A}_k - \frac{1}{2\tau_k g_{i,i}} E, \quad \hat{A}_k = A - BB^T X_k E.
\end{aligned}
\tag{6.43}
$$

We start the investigations at $\mathcal{R}(\sum_{\ell=1}^{j} \mathbf{g}_{j,\ell} Y_{k-1,\ell})$. For that, first consider the sum of Riccati operators

$$\sum_{j=1}^{s} a_{i,j}\mathcal{R}(X_{k-1,j}) = \sum_{j=1}^{s} a_{i,j}\left(C^TC + A^TX_{k-1,j}E + E^TX_{k-1,j}A - E^TX_{k-1,j}BB^TX_{k-1,j}E\right).$$

Further, recall the definitions $\mathbf{X}_k = (X_{k,i})_{i=1}^{s}$ and $\mathbf{Y}_k = (Y_{k,i})_{i=1}^{s}$. Then, from $A_k = A$ being constant and motivated by (5.34), for the linear part, we find

$$\begin{aligned} &\sum_{j=1}^{s} a_{i,j}A^TX_{k-1,j}E + \sum_{j=1}^{s} a_{i,j}E^TX_{k-1,j}A, \ i = 1,\dots,s \\ \Leftrightarrow \quad &((a_{i,j}) \otimes A^T)\mathbf{X}_{k-1}E + ((a_{i,j}) \otimes E^T)\mathbf{X}_{k-1}A. \end{aligned} \tag{6.44}$$

Moreover, from the definition (5.33) of $\mathbf{X}_k$ in terms of the auxiliary variables $\mathbf{Y}_k$ the following reformulation holds:

$$\begin{aligned} &((a_{i,j}) \otimes A^T)\mathbf{X}_{k-1}E+((a_{i,j}) \otimes E^T)\mathbf{X}_{k-1}A \\ &= ((a_{i,j}) \otimes A^T)(G^{-1} \otimes I)\mathbf{Y}_{k-1}E + ((a_{i,j}) \otimes E^T)(G^{-1} \otimes I)\mathbf{Y}_{k-1}A \\ &= ((a_{i,j})G^{-1} \otimes A^T)\mathbf{Y}_{k-1}E + ((a_{i,j})G^{-1} \otimes E^T)\mathbf{Y}_{k-1}A \\ &= ((\mathbf{a}_{i,j}) \otimes A^T)\mathbf{Y}_{k-1}E + ((\mathbf{a}_{i,j}) \otimes E^T)\mathbf{Y}_{k-1}A. \end{aligned} \tag{6.45}$$

with $(\mathbf{a}_{i,j}) = (a_{i,j})G^{-1}$ from (5.35). Then, together with

$$\begin{aligned} &((\mathbf{a}_{i,j}) \otimes A^T)\mathbf{Y}_{k-1}E + ((\mathbf{a}_{i,j}) \otimes E^T)\mathbf{Y}_{k-1}A \\ \Leftrightarrow &\sum_{j=1}^{s} \mathbf{a}_{i,j}A^TY_{k-1,j}E + \sum_{j=1}^{s} \mathbf{a}_{i,j}E^TY_{k-1,j}A, \ i = 1,\dots,s, \end{aligned} \tag{6.46}$$

the sum of Riccati operators $\mathcal{R}(X_{k-1,j})$ can be written in the mixed form

$$\begin{aligned} \sum_{j=1}^{s} a_{i,j}\mathcal{R}(X_{k-1,j}) &= \sum_{j=1}^{s} a_{i,j}\left(C^TC + A^TX_{k-1,j}E + E^TX_{k-1,j}A - E^TX_{k-1,j}BB^TX_{k-1,j}E\right) \\ &= \sum_{j=1}^{s} a_{i,j}\left(C^TC - E^TX_{k-1,j}BB^TX_{k-1,j}E\right) + \sum_{j=1}^{s} \mathbf{a}_{i,j}\left(A^TY_{k-1,j}E + E^TY_{k-1,j}A\right). \end{aligned}$$

Note that in this formulation only the quadratic term of the Riccati operator uses the original variables and analogously to the right hand side $\tilde{W}_{k,i}$ in (6.40), for an autonomous GDRE, we obtain

$$\begin{aligned}\tilde{W}_{k,i} = &- \sum_{j=1}^{s} \Big(\breve{A}_{k,i,j}^T Y_{k-1,j} E + E^T Y_{k-1,j} \breve{A}_{k,i,j} - \mathbf{a}_{i,j} \left(E^T X_k B B^T Y_{k-1,j} E + E^T Y_{k-1,j} B B^T X_k E \right) \Big) \\ &+ \sum_{j=1}^{s} a_{i,j} \left(C^T C - E^T X_{k-1,j} B B^T X_{k-1,j} E \right) + \sum_{j=1}^{s} \mathbf{a}_{i,j} \left(A^T Y_{k-1,j} E + E^T Y_{k-1,j} A \right) \\ &- \sum_{j=1}^{i-1} \frac{\mathbf{g}_{i,j}}{\tau_k} E^T Y_{k,j} E\end{aligned}$$

with $\breve{A}_{k,i,j} = \mathbf{a}_{i,j} A - \frac{\mathbf{b}_{i,j}}{2\tau_k} E$. Now, combining the expressions that are linear in $Y_{k-1,j}$, as well as the quadratic terms containing $E^T X_k B$ and again paying particular attention to $j = s$ with $X_{k-1,s} = X_k$, $Y_{k-1,s} = Y_k$, the right hand side reads

$$\begin{aligned}\tilde{W}_{k,i} &= \sum_{j=1}^{s} a_{i,j} C^T C - \sum_{j=1}^{s-1} a_{i,j} E^T X_{k-1,j} B B^T X_{k-1,j} E + E^T X_k B K_{k,i}^T + K_{k,i} B^T X_k E \\ &\quad + \sum_{j=1}^{s} \frac{\mathbf{b}_{i,j}}{\tau_k} E^T Y_{k-1,j} E - \sum_{j=1}^{i-1} \frac{\mathbf{g}_{i,j}}{\tau_k} E^T Y_{k,j} E, \\ K_{k,i} &= E^T \left(\sum_{j=1}^{s} \mathbf{a}_{i,j} Y_{k-1,j} - \frac{a_{i,s}}{2} X_k \right) B.\end{aligned} \tag{6.47}$$

Thus, the resulting low-rank factor $G_{k,i}$, defining the factorization $\tilde{W}_{k,i} = G_{k,i} G_{k,i}^T$, can be written in the form

$$\begin{aligned}G_{k,i} = \Big[&\sqrt{\sum_{j=1}^{s} a_{i,j}} C^T,\ \jmath \sqrt{a_{i,1}} E^T X_{k-1,1} B, \ldots, \jmath \sqrt{a_{i,s-1}} E^T X_{k-1,s-1} B, \\ &E^T X_k B + K_{k,i},\ \jmath E^T X_k B,\ \jmath K_{k,i}, \\ &\sqrt{\frac{\mathbf{b}_{i,1}}{\tau_k}} E^T \hat{Z}_{k-1,1}, \ldots, \sqrt{\frac{\mathbf{b}_{i,s}}{\tau_k}} E^T \hat{Z}_{k-1,s},\ \jmath \sqrt{\frac{\mathbf{g}_{i,1}}{\tau_k}} E^T \hat{Z}_{k,1}, \ldots, \jmath \sqrt{\frac{\mathbf{g}_{i,i-1}}{\tau_k}} E^T \hat{Z}_{k,i-1}, \Big].\end{aligned} \tag{6.48}$$

Then, the column size of $G_{k,i}$ for autonomous systems is

$$q + \sum_{j=1}^{s-1} m + 3m + \sum_{j=1}^{s} n_{\hat{Z}_{k-1,j}} + \sum_{j=1}^{i-1} n_{\hat{Z}_{k,j}} = \sum_{j=1}^{i-1} n_{\hat{Z}_{k,j}} + \sum_{j=1}^{s} n_{\hat{Z}_{k-1,j}} + q + (s+2)m. \tag{6.49}$$

Note that also for the autonomous case the auxiliary, as well as the original variables are present in the formulations. Still, the original variables only appear in the quadratic terms that, considering the low-rank representations, enter the factors within blocks of column size m. That is, given only a few inputs to the associated linear system that is one of the basic assumptions anyway, the actual size of the solution factors does not directly influence the size of the right hand side factorization and thus is not dramatically affecting the computational effort. Moreover, the reformulations (6.44)-(6.46) allow us to perform all the combinations of the linear expressions, as it was possible for the original Rosenbrock-type peer methods. Therefore, again assuming the column sizes $n_{\mathcal{Z}_{k-1,j}}$ and $n_{\check{\mathcal{Z}}_{k-1,j}}$ of the low-rank factors of $X_{k-1,j}$ and $Y_{k-1,j}$ being comparable, from (6.38) and (6.49), one can easily see that the modified scheme yields less column blocks and is therefore expected to be the more efficient realization.

Classical low-rank Rosenbrock-type Peer scheme for GDLEs Referring to the derivations performed for the GDRE case, the Rosenbrock-type peer scheme applied to the GDLE (6.2) results in the solution of the GALEs

$$\begin{aligned} &\tilde{A}_{k,i}X_{k,i}E^T + EX_{k,i}\tilde{A}_{k,i}^T = -\tilde{W}_{k,i}, \quad i = 1,\dots,s, \\ &\tilde{W}_{k,i} = \tau_k \sum_{j=1}^{i-1} g_{i,j}\left(A_kX_{k,j}E^T + EX_{k,j}A_k^T\right) + \tau_k a_{i,s}B_kB_k^T + b_{i,s}EX_kE^T \\ &\qquad + \sum_{j=1}^{s-1}\left(\tau_k a_{i,j}B_{k-1,j}B_{k-1,j}^T + \check{A}_{k,i,j}^T X_{k-1,j}E + E^TX_{k-1,j}\check{A}_{k,i,j}\right), \end{aligned} \tag{6.50}$$

where $\tilde{A}_{k,i} = \tau_k g_{i,i}A_k - \frac{1}{2}E$ and $\check{A}_{k,i,j} = \tau_k a_{i,j}(A_{k-1,j} - A_k) + \frac{b_{i,j}}{2}E$, which is identical to the GDRE case. Defining the factors

$$\begin{aligned} \mathcal{Z}_{k,j} &= \left[A_kZ_{k,j} + EZ_{k,j},\ \jmath A_kZ_{k,j},\ \jmath EZ_{k,j}\right] \in \mathbb{C}^{n\times 3n_{Z_{k,j}}}, \\ \check{\mathcal{Z}}_{k,i,j} &= \left[\ \sqrt{\tau_k a_{i,j}}B_{k-1,j}\ \check{A}_{k,i,j}Z_{k-1,j} + EZ_{k-1,j},\ \jmath\check{A}_{k,i,j}Z_{k-1,j},\ \jmath EZ_{k-1,j}\right] \in \mathbb{C}^{n\times(m+3n_{Z_{k-1,j}})}, \end{aligned}$$

for the Lyapunov-type expressions, the corresponding expression $\tilde{W}_{k,i} = G_{k,i}G_{k,i}^T$ is defined by the factor

$$G_{k,i} = \left[\ \sqrt{\tau_k g_{i,1}}\mathcal{Z}_{k,1}, \dots, \sqrt{\tau_k g_{i,i-1}}\mathcal{Z}_{k,s}, \sqrt{\tau_k a_{i,s}}B_k, \sqrt{b_{i,s}}EZ_k, \check{\mathcal{Z}}_{k,i,1}, \dots, \check{\mathcal{Z}}_{k,i,s-1}\right] \tag{6.51}$$

with a column size of

$$3\sum_{j=1}^{i-1} n_{Z_{k,j}} + 3\sum_{j=1}^{s-1} n_{Z_{k-1,j}} + n_{Z_{k-1,s}} + sm.$$

Again, following the statements for the GDRE, in the autonomous GDLE case, we have to solve the GALE

$$\begin{aligned} &\tilde{A}_{k,i}X_{k,i}E^T + EX_{k,i}\tilde{A}_{k,i}^T = -\tilde{W}_{k,i} \quad i = 1,\dots,s \\ &\tilde{W}_{k,i} = \tau_k \sum_{j=1}^{i-1} g_{i,j}\left(AX_{k,j}E^T + EX_{k,j}A^T\right) + \sum_{j=1}^{s}\left(\tau_k a_{i,j}BB^T + b_{i,j}EX_{k-1,j}E^T\right), \end{aligned} \tag{6.52}$$

with $\tilde{A}_{k,i} = \tau_k g_{i,i} A - \frac{1}{2}E$ and the analog right hand side factorization to (6.51) is given by

$$\begin{aligned} G_{k,i} = \Big[&\sqrt{\tau_k g_{i,1}} \mathcal{Z}_{k,1}, \ldots, \sqrt{\tau_k g_{i,i-1}} \mathcal{Z}_{k,s}, \\ &\sqrt{\tau_k \sum_{j=1}^{s} a_{i,j}} B, \; \sqrt{b_{i,1}} E Z_{k-1,1}, \ldots, \sqrt{b_{i,s}} E Z_{k-1,s} \Big] \end{aligned} \tag{6.53}$$

with $\mathcal{Z}_{k,j} = [AZ_{k,j} + EZ_{k,j}, \; \jmath A Z_{k,j}, \; \jmath E Z_{k,j}]$ and a total column size of

$$3 \sum_{j=1}^{i-1} n_{Z_{k,j}} + \sum_{j=1}^{s} n_{Z_{k-1,j}} + m.$$

For the application of the modified Rosenbrock-type peer method (5.38) to the GDLE (6.2), we skip the detailed modifications, performed for the GDRE. Finally, we have to solve the GALEs

$$\begin{aligned} \tilde{A}_{k,i} Y_{k,i} E^T + E Y_{k,i} \tilde{A}_{k,i}^T &= -\tilde{W}_{k,i}, \quad i = 1, \ldots, s, \\ \tilde{W}_{k,i} &= -\sum_{j=1}^{s} \Big(\breve{A}_{k,i,j}^T Y_{k-1,j} E + E^T Y_{k-1,j} \breve{A}_{k,i,j} \Big) \\ &\quad + \tau_k \sum_{j=1}^{s} a_{i,j} \mathcal{L}(t_{k-1,j}, X_{k-1,j}) - \sum_{j=1}^{i-1} \frac{\mathbf{g}_{i,j}}{\tau_k} E Y_{k,j} E^T \end{aligned} \tag{6.54}$$

with $\breve{A}_{k,i,j} = \mathbf{a}_{i,j} A_k - \frac{\mathbf{b}_{i,j}}{2\tau_k} E$, that again is given in terms of the original and auxiliary variables. Together with the expression

$$\breve{\mathcal{Z}}_{k-1,i,j} = \Big[\Big(\breve{A}_{k,i,j} \hat{Z}_{k-1,j} + E \hat{Z}_{k-1,j} \Big), \jmath \breve{A}_{k,i,j} \hat{Z}_{k-1,j}, \; \jmath \sqrt{\mathbf{a}_{i,j}} E \hat{Z}_{k-1,j} \Big] \in \mathbb{C}^{n \times (3 n_{\hat{Z}_{k-1,j}})},$$

representing the linear part $\breve{A}_{k,i,j}^T Y_{k-1,j} E + E^T Y_{k-1,j} \breve{A}_{k,i,j}$, and the decomposition

$$\begin{aligned} &\mathcal{L}(t_{k-1,j}, X_{k-1,j}) = \mathcal{Z}_{k-1,j} \mathcal{Z}_{k-1,j}^T, \\ &\mathcal{Z}_{k-1,j} = \Big[B_{k-1,j}, \; A_{k-1,j} Z_{k-1,j} + E Z_{k-1,j}, \; \jmath A_{k-1,j} Z_{k-1,j}, \; \jmath E Z_{k-1,j} \Big] \in \mathbb{C}^{n \times (m + 3 n_{Z_{k-1,j}})}, \end{aligned}$$

this results in the right hand side factor

$$\begin{aligned} G_{k,i} = \Big[&\jmath \breve{\mathcal{Z}}_{k-1,i,1}, \ldots, \jmath \breve{\mathcal{Z}}_{k-1,i,s}, \sqrt{\tau_k a_{i,1}} \mathcal{Z}_{k-1,1}, \ldots, \sqrt{\tau_k a_{i,s}} \mathcal{Z}_{k-1,s}, \\ &\jmath \sqrt{\frac{\mathbf{g}_{i,1}}{\tau_k}} E \hat{Z}_{k,1}, \ldots, \jmath \sqrt{\frac{\mathbf{g}_{i,i-1}}{\tau_k}} E \hat{Z}_{k,i-1}, \Big], \end{aligned}$$

that is of column size

$$\sum_{j=1}^{i-1} n_{\hat{Z}_{k,j}} + 3 \sum_{j=1}^{s} \Big(n_{\hat{Z}_{k-1,j}} + n_{Z_{k-1,j}} \Big) + sm.$$

Here, we face the same situation as for the DRE case. That is, avoiding the application of the Jacobian to the current solutions cannot compensate the fact that the combination of the linear parts and the function evaluations is not feasible due to the use of the mixed formulation.

Following the manipulations to derive (6.47) from (6.43) for the GDRE, in the autonomous case the Rosenbrock-type peer method yields the GALEs

$$\begin{aligned} \tilde{A}_{k,i} Y_{k,i} E^T + E Y_{k,i} \tilde{A}_{k,i}^T &= -\left(\tau_k \sum_{j=1}^{s} a_{i,j} B B^T + \sum_{j=1}^{s} \frac{\mathbf{b}_{i,j}}{\tau_k} E Y_{k-1,j} E^T - \sum_{j=1}^{i-1} \frac{\mathbf{g}_{i,j}}{\tau_k} E Y_{k,j} E^T \right), \\ \tilde{A}_{k,i} &= A - \frac{1}{2\tau_k g_{i,i}} E. \end{aligned} \tag{6.55}$$

The right hand side factor $G_{k,i}$ defining the above right hand side $\tilde{W}_{k,i}$ can then be written in the form

$$G_{k,i} = \left[\sqrt{\sum_{j=1}^{s} a_{i,j}} B,\ \frac{\mathbf{b}_{i,1}}{\tau_k} E\hat{Z}_{k-1,1}, \ldots, \frac{\mathbf{b}_{i,s}}{\tau_k} E\hat{Z}_{k-1,s}\ \jmath \frac{\mathbf{g}_{i,1}}{\tau_k} E\hat{Z}_{k,1}, \ldots, \jmath \frac{\mathbf{g}_{i,i-1}}{\tau_k} E\hat{Z}_{k,i-1}, \right], \tag{6.56}$$

and is of column size

$$\sum_{j=1}^{i-1} n_{\hat{Z}_{k,j}} + \sum_{j=1}^{s} n_{\hat{Z}_{k-1,j}} + m.$$

6.1.6. Limitation of the Classical Low-Rank Factorization

In Section 2.1.4, we have seen that both, the ADI and Krylov subspace methods, are very well suited for solving large-scale Lyapunov equations with real right hand side factorizations $W = GG^T$, $G \in \mathbb{R}^{n \times n_G}$. Still, it has been presented that for the above integration methods of order $s \geq 2$ the right hand sides $W = W^T \in \mathbb{R}^{n \times n}$ of the Lyapunov equations to be solved become indefinite and their classical low-rank factorizations lead to expressions $W = GG^T$ with $G \in \mathbb{C}^{n \times n_G}$, being complex. Moreover, the splitting of the right hand sides into separate definite parts was observed to fail for the application to the matrix-valued ODE solvers. Therefore, as a representative for the GALEs inside the integration methods, here we consider the GALE

$$F^T X E + E^T X F = -W = -GG^T,\ G \in \mathbb{C}^{n \times n_G}. \tag{6.57}$$

Then, still being interested in using the ADI iteration and Krylov subspace methods based on the classical two-term factorization, the procedures need to be investigated for the special case $G \in \mathbb{C}^{n \times n_G}$. Further, up to a re-ordering of the column blocks, the affected right hand side factors are given in a partitioning

$$G = \begin{bmatrix} G_1, & \jmath G_2 \end{bmatrix} \in \mathbb{C}^{n \times n_G}, \quad G_1 \in \mathbb{R}^{n \times n_{G_1}},\ G_2 \in \mathbb{R}^{n \times n_{G_2}},\ n_{G_1} + n_{G_2} = n_G \tag{6.58}$$

with purely real and imaginary blocks G_1 and $\jmath G_2$. Since now, G is already complex, the realification ideas for the ADI iteration, mentioned in Section 2.1, are no longer meaningful in its present fashion. Thus, our considerations for the ADI fall back to the original sequence with efficient low-rank residual computation (see [126, Algorithm 3.2] with the notation inherited from Section 2.1), given by

$$\begin{aligned} V_j &= (F + \alpha_j E)^{-1} R_{j-1}, \; R_0 = G, \\ R_j &= R_{j-1} - 2\,\mathrm{Re}\big(\alpha_j\big)\, E V_j, \\ Z_j &= \Big[Z_{j-1}, \; \sqrt{-2\,\mathrm{Re}\big(\alpha_j\big)} V_j \Big]. \end{aligned}$$

If now, the initial shift $\alpha_1 \in \mathbb{R}$, the special partitioning of G carries over to V_1. That is, V_1 is of the form $V_1 = [\tilde{V}_1, \; \jmath \hat{V}_1] = [S_1 G_1, \; \jmath S_1 G_2]$ with $S_1 = (F + \alpha_1 E)^{-1} \in \mathbb{R}^{n \times n}$ and real matrices $\tilde{V}_1$ and $\hat{V}_2$. Moreover the block sizes of the pure real and imaginary parts are preserved. Since all further steps are linear in the iterates, the partitioning is also transferred to $R_1 = [\tilde{R}_1, \; \jmath \hat{R}_1]$, $Z_1 = [\tilde{Z}_1, \; \jmath \hat{Z}_1]$ and the subsequent step $j = 2$ starts with the same situation. This implies that as long as the set of shifts is real, the structure (6.58) is preserved and the final solution X to Equation (6.57) can be recovered from the real inner product $ZZ^T = Z_1 Z_1^T - Z_2 Z_2^T$, where $Z = [Z_1, \; \jmath Z_2]$ denotes the final solution iterate of the ADI procedure. Note that the structure of ZZ^T is similar to the superposition ansatz briefly mentioned in Equation (6.9) in Section 6.1.1. However, if one of the shifts α_j becomes complex, the same holds for S_j, and for the iterate V_j, we obtain

$$V_j = [S_j \tilde{R}_{j-1}, \jmath S_j \hat{R}_{j-1}] = [\tilde{V}_j, \jmath \hat{V}_j]$$

with the single blocks $\tilde{V}_j$ and $\hat{V}_j$ both being complex. That is, the proper partitioning into pure real and imaginary parts and thus the special structure (6.58) is lost. Therefore, in general, the solution X has to be setup in the form ZZ^H, where, again, $Z \in \mathbb{C}^{n \times n_Z}$ is the final low-rank solution factor, obtained from the ADI iteration. Assuming the complex shifts to appear in complex conjugate pairs, although Z is complex, it can even be guaranteed that for the solution X, we have $X \in \mathbb{R}^{n \times n}$. So far, this seems not to be a particular problem. Still, note that within the integration schemes, the solution X_k enters the subsequent right hand side $W_{k+1} \in \mathbb{R}^{n \times n}$. For simplicity, here we consider the illustrative right hand side

$$W_{k+1} = H_1 H_1^T - H_2 H_2^T + X_k$$

that is only intended to capture the basic features of the actual right hand sides arising within the time integration methods. Here, H_1, H_2 are assumed to be real and of low numerical rank, and $X_k \approx Z_k Z_k^H$, where Z_k originates from the ADI iteration performed with complex shifts. That is, the required factorization of the form $G_{k+1} G_{k+1}^T$ cannot be constructed due to the mixture of real and complex inner products in the summands forming the right hand side. Now, one may ask for a decomposition of the form $G_{k+1} G_{k+1}^H$ in the first place. Since the right hand side at hand is real symmetric but indefinite, the answer is quite simple. There is no decomposition of the desired form. For conviction, assume $W = W^T \in \mathbb{R}^{n\times n}$ and indefinite. Further assume, there is a decomposition of the form $W = GG^H$ with G being complex. Then, G can be written as $G = G_1 + \jmath G_2$, where G_1, G_2 are real. Then, we obtain

$$\begin{aligned} GG^H &= (G_1 + \jmath G_2)(G_1 + \jmath G_2)^H \\ &= (G_1 + \jmath G_2)(G_1^T - \jmath G_2^T) \\ &= G_1G_1^T - \jmath G_1G_2^T + \jmath G_2G_1^T + G_2G_2^T \end{aligned}$$

that can only become real if $G_1G_2^T$ is symmetric or, $G_1 = 0$ or $G_2 = 0$, meaning the factor G is purely real or imaginary, respectively. Thus, in general, GG^H is complex and therefore contradicts the assumption for W. That, in a final consequence requires a mixed LRCF representation, resulting in a considerable implementational effort and is thus not recommended.

In case of the LRCF Krylov subspace Lyapunov equation solver, the main step is to setup the corresponding Krylov subspace as given in Step 1 of Algorithm 2.3. That is, the right hand side factor G from (6.58) underlies a repeated application of shifted linear system solves. From the statements above, we can easily conclude that for real shifts, the structure is preserved and the real inner product ZZ^T is valid, whereas the appearance of complex shifts, again, leads to the loss of structure and therefore requires special attention while forming the solution X. Recall that explicitly forming X is rather theoretical and not recommended for practical use.

6.2. Symmetric Indefinite Low-Rank Factorization

In Section 6.1, we have seen that all implicit time integration methods of order $s \geq 2$ lead to indefinite right hand sides of the ALEs and therefore give rise to complex factorizations GG^T with $G \in \mathbb{C}^{n\times n_G}$, resulting in the demand for complex storage and arithmetic. If in addition complex shifts occur within the Lyapunov solution methods the corresponding algorithms can no longer be applied straightforwardly. Introducing a factorization, that is more natural for symmetric indefinite matrices, of the form GSG^T with $G \in \mathbb{R}^{n\times n_G}$ and $S = S^T \in \mathbb{R}^{n_G\times n_G}$, $n_G \ll n$ to the right hand sides, the resulting GALEs become

$$A^TXE + E^TXA = -GSG^T,$$

avoiding complex data. Moreover re-arranging the occurring blocks within the factors G, S, allows us to eliminate redundant information and therefore to reduce the overall column size n_G of the right hand side factor G, compared to the classical low-rank strategy. Similar to the classical low-rank formulation, in the remainder of this section, we assume the solution X_j to admit a factorization $X_j = L_jD_jL_j^T$ with $L_j \in \mathbb{R}^{n\times n_{L_j}}$ and $D_j = D_j^T \in \mathbb{R}^{n_{L_j}\times n_{L_j}}$ at all instances t_j of the discrete time interval to be considered. Note that from the theory it is well-known that the numerical computation of a symmetric indefinite factorization of a matrix is in general unstable that can be circumvented by a suitable application of permutation matrices, see e.g., [84, Section 4.4]. Still, this does not affect the following considerations. Here, the special structure of the GALEs allows to analytically determine a symmetric indefinite formulation of the right hand sides, which is directly transferred to the solutions by the solution strategies. That is, the symmetric indefinite factorization is not performed explicitly.

In the following subsections the specific symmetric indefinite factorizations and the a-priori analytically applied column compression are explained in detail for the integration schemes presented in Chapter 5.

6.2.1. Backward Differentiation Formulas

Low-rank symmetric indefinite BDF scheme for GDREs Again, the GALE, originating from the application of Newton's method to the GARE (6.4), arising in the BDF scheme, has the form

$$\begin{aligned}\hat{A}_{k+1}^{(\ell)^T} X_{k+1}^{(\ell)} E + E^T X_{k+1}^{(\ell)} \hat{A}_{k+1}^{(\ell)} &= -\left(\tau_k \beta C_{k+1}^T C_{k+1} - E^T \left(\sum_{j=1}^{s} \alpha_j X_{k+1-j}\right) E\right) \\ &\quad - \tau_k \beta E^T X_{k+1}^{(\ell-1)} B_{k+1} B_{k+1}^T X_{k+1}^{(\ell-1)} E.\end{aligned}$$

Then, a symmetric indefinite factorization $-G_{k+1}^{(\ell)} S_{k+1}^{(\ell)} G_{k+1}^{(\ell)^T}$ of the right hand side reads

$$\begin{aligned}G_{k+1}^{(\ell)} &= \left[C_{k+1}^T, \quad E^T L_k, \ldots, E^T L_{k+1-s}, \quad E^T L_{k+1}^{(\ell-1)} D_{k+1}^{(\ell-1)} L_{k+1}^{(\ell-1)^T} B_{k+1}\right] \in \mathbb{R}^{n \times (q + \sum_{j=1}^{s} n_{L_j} + m)}, \\ S_{k+1}^{(\ell)} &= \begin{bmatrix} \tau_k \beta I_q & & & & \\ & -\alpha_1 D_k & & & \\ & & \ddots & & \\ & & & -\alpha_s D_{k+1-s} & \\ & & & & \tau_k \beta I_m \end{bmatrix} \in \mathbb{R}^{(q + \sum_{j=1}^{s} n_{L_j} + m) \times (q + \sum_{j=1}^{s} n_{L_j} + m)}\end{aligned} \tag{6.59}$$

with I_q and I_m being identity matrices of appropriate size m and q, respectively. Here, the indefiniteness is captured by the middle term S_{k+1}. That is, the square roots of the negative coefficients α_j, giving rise to complex data, are no longer present such that the factors G_{k+1} and S_{k+1} stay real over the entire integration process.

Low-rank symmetric indefinite BDF scheme for GDLEs Analogously, in case of the GDLE (6.2), we obtain the GALEs

$$\tilde{A}_{k+1} X_{k+1} E^T + E X_{k+1} \tilde{A}_{k+1} = -G_{k+1} S_{k+1} G_{k+1}^T,$$

where the symmetric indefinite right hand side factorization becomes

$$
\begin{aligned}
G_{k+1} &= \begin{bmatrix} B_{k+1}, & EL_k, \dots, EL_{k+1-s}, \end{bmatrix} \in \mathbb{R}^{n\times(m+\sum_{j=1}^{s} n_{L_j})}, \\
S_{k+1} &= \begin{bmatrix} \tau_k\beta I_m & & & \\ & -\alpha_1 D_k & & \\ & & \ddots & \\ & & & -\alpha_s D_{k+1-s} \end{bmatrix} \in \mathbb{R}^{(m+\sum_{j=1}^{s} n_{L_j})\times(m+\sum_{j=1}^{s} n_{L_j})}.
\end{aligned} \tag{6.60}
$$

6.2.2. Rosenbrock Methods

Low-rank symmetric indefinite Rosenbrock methods for GDREs Note that the GALE (6.10) has a definite right hand side and therefore already admits a real formulation, using the classical low-rank approach.

Therefore, a naively determined factorization $G_kS_kG_k^T$ of the right hand side with factors

$$
\begin{aligned}
G_k &= \begin{bmatrix} C_k^T, & E^TL_k, & E^TL_k \end{bmatrix} & \in \mathbb{R}^{n\times(q+2n_{L_k})} \\
S_k &= \begin{bmatrix} I_q & & \\ & D_kL_k^TB_kB_kL_kD_k & \\ & & \frac{1}{\tau_k}D_k \end{bmatrix} & \in \mathbb{R}^{(q+2n_{L_k})\times(q+2n_{L_k})}
\end{aligned} \tag{6.61}
$$

cannot improve the numerical computations. Still, the factorization (6.61) reveals the redundant information, hidden in the factorization of the right hand side. Hence, re-arranging the factors in the form

$$
\begin{aligned}
G_k &= \begin{bmatrix} C_k^T, & E^TL_k \end{bmatrix} & \in \mathbb{R}^{n\times(q+n_{L_k})}, \\
S_k &= \begin{bmatrix} I_q & \\ & D_kL_k^TB_kB_kL_kD_k + \frac{1}{\tau_k}D_k \end{bmatrix} & \in \mathbb{R}^{(q+n_{L_k})\times(q+n_{L_k})}
\end{aligned} \tag{6.62}
$$

allows us to eliminate the redundant blocks in G_k. That is, the re-arrangement reduces the column size of the factor G_k by n_{L_k}.

In order to analyze the Ros(2) scheme (6.11), we again start with the factorization of the Riccati operator $\mathcal{R}(t_k, X_k)$, which is the negative right hand side of the first-stage Equation (6.11b) of the scheme. As for the classical low-rank case, the linear term is of particular interest and its symmetric indefinite factorization can be written in the form

$$
A_k^TL_kD_kL_k^TE + E^TL_KD_kL_k^TA_k = \begin{bmatrix} A_k^TL_k, & E^TL_k \end{bmatrix} \begin{bmatrix} & D_k \\ D_k & \end{bmatrix} \begin{bmatrix} L_k^TA_k \\ L_k^TE \end{bmatrix},
$$

where the middle block $\begin{bmatrix} & D_k \\ D_k & \end{bmatrix}$ realizes the block-wise exchange of the columns, if applied from the right and the swapping of the row blocks, applied from the left.

Finally, a straight forward symmetric indefinite factorization $G_k S_k G_k^T$ of

$$\mathcal{R}(t_k, X_k) = C_k^T C_k + A_k^T L_k D_k L_k^T E + E^T L_k D_k L_k^T A_k - E^T L_k D_k L_k^T B_k B_k^T L_k D_k L_k^T E$$

is given by the factors

$$\begin{aligned} G_k^{(1)} &= \begin{bmatrix} C_k^T, & A_k^T L_k, & E^T L_k, & E^T L_k \end{bmatrix} \in \mathbb{R}^{n \times (q+3n_{L_k})}, \\ S_k^{(1)} &= \begin{bmatrix} I_q & & & \\ & & D_k & \\ & D_k & & \\ & & & -D_k L_k^T B_k B_k^T L_k D_k \end{bmatrix} \in \mathbb{R}^{(q+3n_{L_k}) \times (q+3n_{L_k})}, \end{aligned} \tag{6.63}$$

whereas a more sophisticated representation based on a re-arrangement, similar to the first-order method, results in

$$\begin{aligned} G_k^{(1)} &= \begin{bmatrix} C_k^T, & A_k^T L_k, & E^T L_k \end{bmatrix} \in \mathbb{R}^{n \times (q+2n_{L_k})}, \\ S_k^{(1)} &= \begin{bmatrix} I_q & & \\ & & D_k \\ & D_k & -D_k L_k^T B_k B_k^T L_k D_k \end{bmatrix} \in \mathbb{R}^{(q+2n_{L_k}) \times (q+2n_{L_k})}. \end{aligned} \tag{6.64}$$

Consequently, the re-arrangement saves n_{L_k} columns in the right hand side factor G_k of the first-stage equation of the Ros(2) procedure. Given the solution $K_1 \approx T_1 M_1 T_1^T$ with $T_1 \in \mathbb{R}^{n \times n_{T_1}}$, $M_1 \in \mathbb{R}^{n_{T_1} \times n_{T_1}}$ to the first-stage equation, using (6.14), the right hand side of the second-stage Equation (6.11c), in terms of the symmetric indefinite factorization becomes

$$G_k^{(2)} = \begin{bmatrix} C_{k+1}^T, & A_{k+1}^T L_k, & E^T L_k, & E^T L_k, & A_{k+1}^T T_1, & E^T T_1, & E^T T_1, & E^T L_k, & E^T T_1, & E^T T_1 \end{bmatrix}$$

$$S_k^{(2)} = \begin{bmatrix} I_q & & & & & & & & & \\ & & D_k & & & & & & & \\ & D_k & & & & & & & & \\ & & & -D_k L_k^T B_{k+1} B_{k+1}^T L_k^T D_k & & & & & & \\ & & & & & \tau_k M_1 & & & & \\ & & & & \tau_k M_1 & & & & & \\ & & & & & & -\tau_k^2 M_1 T_1 B_{k+1} B_{k+1}^T T_1^T M_1 & & & \\ & & & & & & & \ddots & -\tau_k D_k L_k^T B_{k+1} B_{k+1}^T T_1 M_1 & \\ & & & & & & & -\tau_k M_1 T_1^T B_{k+1} B_{k+1}^T L_k D_k & & \\ & & & & & & & & & -2M_1 \end{bmatrix}$$

with $G_k^{(2)} \in \mathbb{R}^{n \times (q+4n_{L_k}+5n_{T_1})}$ and $S_k^{(2)} \in \mathbb{R}^{(q+4n_{L_k}+5n_{T_1}) \times (q+4n_{L_k}+5n_{T_1})}$.

Eliminating all the repetitively appearing blocks in $G_k^{(2)}$, the re-arranged versions of the right hand side factors are given by

$$G_k^{(2)} = \left[C_{k+1}^T, \ A_{k+1}^T L_k, \ E^T L_k, \ A_{k+1}^T T_1, \ E^T T_1\right]$$

$$S_k^{(2)} = \begin{bmatrix} I_q & & & & \\ & & D_k & & \\ & D_k & -D_k L_k^T B_{k+1} B_{k+1}^T L_k^T D_k & & \tau_k D_k L_k^T B_{k+1} B_{k+1}^T T_1 M_1 \\ & & & & \tau_k M_1 \\ & & \tau_k M_1 T_1^T B_{k+1} B_{k+1}^T L_k D_k & \tau_k M_1 & -\tau_k^2 M_1 T_1 B_{k+1} B_{k+1}^T T_1^T M_1 - 2M_1 \end{bmatrix},$$

where $G_k^{(2)} \in \mathbb{R}^{n \times (q+2n_{L_k}+2n_{T_1})}$ and $S_k^{(2)} \in \mathbb{R}^{(q+2n_{L_k}+2n_{T_1}) \times (q+2n_{L_k}+2n_{T_1})}$. That is, a saving of $2n_{L_k}+3n_{T_1}$ columns could be achieved for the second-stage equation. In total $3n_{L_k}+3n_{T_1}$ columns for the entire second-order Rosenbrock method are eliminated.

Considering the Ros(2) scheme (6.15) for autonomous GDREs, the symmetric indefinite formulation of the first-stage equation does not change except for the constant system matrices and is given by

$$G_k^{(1)} = \left[C^T, \ A^T L_k, \ E^T L_k\right], \qquad S_k^{(1)} = \begin{bmatrix} I_q & & \\ & & D_k \\ & D_k & -D_k L_k^T B_k B_k^T L_k D_k \end{bmatrix}$$

with $G_k^{(1)} \in \mathbb{R}^{n \times (q+2n_{L_k})}$, $S_k^{(1)} \in \mathbb{R}^{(q+2n_{L_k}) \times (q+2n_{L_k})}$.

For the simplified second stage, we obtain

$$G_k^{(21)} = \left[E^T T_1, \ E^T T_1\right], \qquad S_k^{(21)} = \begin{bmatrix} \tau_k^2 M_1 T_1^T B B^T T_1 M_1 & \\ & (2 - \frac{1}{\gamma}) M_1 \end{bmatrix}$$

with $G_k^{(21)} \in \mathbb{R}^{n \times 2n_{T_1}}$, $S_k^{(21)} \in \mathbb{R}^{2n_{T_1} \times 2n_{T_1}}$ and in condensed form

$$G_k^{(21)} = \left[E^T T_1\right], \qquad S_k^{(21)} = \left[\tau_k^2 M_1 T_1^T B B^T T_1 M_1 + (2 - \tfrac{1}{\gamma}) M_1\right]$$

with $G_k^{(21)} \in \mathbb{R}^{n \times n_{T_1}}$ and $S_k^{(21)} \in \mathbb{R}^{n_{T_1} \times n_{T_1}}$. Here, again a reduction of the column size n_{G_k} by n_{T_1} was possible.

Recall from Section 2.1.4 that the number of columns n_{G_k} of the factor G_k is decisive for the performance of the Lyapunov solvers. Using the ADI iteration, n_{G_k} is the number of linear system solves to be performed at every ADI step, see Algorithms 2.1 and 2.2. Regarding the Krylov subspace methods, the number of columns n_{G_k} defines, in addition to the number of system solves, the dimension of the low-dimensional subspace, the Lyapunov equation to be solved is projected on.

Low-rank symmetric indefinite Rosenbrock methods for GDLEs The symmetric indefinite factorization of the right hand side of the first-order scheme (6.16) applied to the non-autonomous GDLE (6.2) yields the factors

$$G_k = \begin{bmatrix} B_k, & EL_k \end{bmatrix} \in \mathbb{R}^{n \times (m+n_{L_k})}, \qquad S_k = \begin{bmatrix} I_m & \\ & \frac{1}{\tau_k} D_k \end{bmatrix} \in \mathbb{R}^{(m+n_{L_k}) \times (m+n_{L_k})}.$$

For the non-autonomous second-order scheme (6.17), the factorizations of right hand sides of the first- and second-stage Lyapunov equations become

$$G_k^{(1)} = \begin{bmatrix} B_k, & A_k L_k, & EL_k, \end{bmatrix} \in \mathbb{R}^{n \times (m+2n_{L_k})}, \qquad S_k^{(1)} = \begin{bmatrix} I_m & & \\ & & D_k \\ & D_k & \end{bmatrix} \in \mathbb{R}^{(m+2n_{L_k}) \times (m+2n_{L_k})},$$

$$G_k^{(2)} = \begin{bmatrix} B_{k+1}, & A_{k+1} L_k, & EL_k, & A_{k+1} T_1, & ET_1 \end{bmatrix} \in \mathbb{R}^{n \times (m+2n_{L_k}+2n_{T_1})},$$

$$S_k^{(2)} = \begin{bmatrix} I_m & & & & \\ & & D_k & & \\ & D_k & & & \\ & & & & M_1 \\ & & & M_1 & -\frac{2}{\tau_k} M_1 \end{bmatrix} \in \mathbb{R}^{n \times (m+2n_{L_k}+2n_{T_1})}$$

and for the autonomous case (6.18), we obtain

$$G_k^{(1)} = \begin{bmatrix} B, & AL_k, & EL_k, \end{bmatrix} \in \mathbb{R}^{n \times (m+2n_{L_k})}, \qquad S_k^{(1)} = \begin{bmatrix} I_m & & \\ & & D_k \\ & D_k & \end{bmatrix} \in \mathbb{R}^{(m+2n_{L_k}) \times (m+2n_{L_k})},$$

$$G_k^{(2)} = ET_1 \in \mathbb{R}^{n \times (n_{T_1})}, \qquad S_k^{(2)} = (2 - \frac{1}{\gamma}) M_1 \in \mathbb{R}^{n_{T_1} \times n_{T_1}}.$$

6.2.3. Midpoint Rule

From the previous subsection, we have seen that re-arranging blocks allows to reduce the overall size of the factor n_{G_k}. Since the formulations are rather longish and the elimination procedure is already sufficiently presented by the given examples of the Rosenbrock methods, in the remainder, we only state the final shortened representation of the symmetric indefinite factorizations.

Low-rank symmetric indefinite Midpoint rule for GDREs The symmetric indefinite factors of the right hand side of the GALE (6.20), arising inside the Midpoint rule, are given as

$$G_k^{(\ell)} = \begin{bmatrix} C_{k'}^T, & \breve{A}_k^T L_k, & E^T L_k, & E^T L_{k+1}^{(\ell-1)} D_{k+1}^{(\ell-1)} L_{k+1}^{(\ell-1)^T} B_{k'} \end{bmatrix},$$

$$S_k^{(\ell)} = \begin{bmatrix} \tau_k I_q & & & \\ & & D_k & \\ & D_k & -\frac{\tau_k}{4} D_K B_{k'} B_{k'}^T L_k D_k & \\ & & & \frac{\tau_k}{4} I_m \end{bmatrix}$$

with $G_k^{(\ell)} \in \mathbb{R}^{n\times(q+2n_{L_k}+m)}$ and $S_k^{(\ell)} \in \mathbb{R}^{(q+2n_{L_k}+m)\times(q+2n_{L_k}+m)}$.

Low-rank symmetric indefinite Midpoint rule for GDLEs Again, consider the GDLE (6.2). For the GALE (6.21) originating from the application of the Midpoint rule, we obtain the symmetric indefinite right hand side factors

$$G_k^{(\ell)} = \begin{bmatrix} B_{k'}, & \tilde{A}_k^T L_k, & E^T L_k \end{bmatrix} \in \mathbb{R}^{n\times(m+2n_{L_k})}, \qquad S_k^{(\ell)} = \begin{bmatrix} \tau_k I_m & & \\ & & D_k \\ & D_k & \end{bmatrix} \in \mathbb{R}^{(m+2n_{L_k})\times(m+2n_{L_k})}.$$

6.2.4. Trapezoidal Rule

Low-rank symmetric indefinite Trapezoidal rule for GDREs The symmetric indefinite factored representation of the GALE within the Trapezoidal rule is defined by the factors

$$\begin{aligned} G_k^{(\ell)} &= \begin{bmatrix} C_{k+1}^T, & C_k^T, & \check{A}_k^T L_k, & E^T L_k, & E^T L_{k+1}^{(\ell-1)} D_{k+1}^{(\ell-1)} L_{k+1}^{(\ell-1)^T} B_{k+1} \end{bmatrix}, \\ S_k^{(\ell)} &= \begin{bmatrix} \frac{\tau_k}{2} I_{2q} & & & \\ & & D_k & \\ & D_k & -\frac{\tau_k}{2} D_k L_k^T B_k B_k^T L_k D_k & \\ & & & \frac{\tau_k}{2} I_m \end{bmatrix} \end{aligned}$$

with $G_k^{(\ell)} \in \mathbb{R}^{n\times(2q+2n_{L_k}+m)}$ and $S_k^{(\ell)} \in \mathbb{R}^{(2q+2n_{L_k}+m)\times(2q+2n_{L_k}+m)}$.

In the autonomous case, we obtain

$$\begin{aligned} G_k^{(\ell)} &= \begin{bmatrix} C^T, & \check{A}^T L_k, & E^T L_k, & E^T L_{k+1}^{(\ell-1)} D_{k+1}^{(\ell-1)} L_{k+1}^{(\ell-1)^T} B \end{bmatrix}, \\ S_k^{(\ell)} &= \begin{bmatrix} \tau_k I_q & & & \\ & & D_k & \\ & D_k & -\frac{\tau_k}{2} D_k L_k^T B B^T L_k D_k & \\ & & & \frac{\tau_k}{2} I_m \end{bmatrix} \end{aligned}$$

with $G_k^{(\ell)} \in \mathbb{R}^{n\times(q+2n_{L_k}+m)}$ and $S_k^{(\ell)} \in \mathbb{R}^{(q+2n_{L_k}+m)\times(q+2n_{L_k}+m)}$.

Low-rank symmetric indefinite Trapezoidal rule for GDLEs For the non-autonomous GDLE, the factors of the associated GALE read

$$G_k^{(\ell)} = \begin{bmatrix} B_{k+1}, & B_k, & \tilde{A}_k^T L_k, & E^T L_k \end{bmatrix} \in \mathbb{R}^{n\times(2m+2n_{L_k})}, \quad S_k^{(\ell)} = \begin{bmatrix} \frac{\tau_k}{2} I_{2m} & & \\ & & D_k \\ & D_k & \end{bmatrix} \in \mathbb{R}^{(2m+2n_{L_k})\times(2m+2n_{L_k})}.$$

Exploiting $B_{k+1} = B_k$ for the autonomous GDLE, the factors become

$$G_k^{(\ell)} = \begin{bmatrix} B, & \tilde{A}^T L_k, & E^T L_k \end{bmatrix} \in \mathbb{R}^{n \times (m+2n_{L_k})}, \qquad S_k^{(\ell)} = \begin{bmatrix} \tau_k I_m & & \\ & & D_k \\ & D_k & \end{bmatrix} \in \mathbb{R}^{(m+2n_{L_k}) \times (m+2n_{L_k})}.$$

6.2.5. Peer Methods

For the implicit and Rosenbrock-type peer methods, we have seen that the classical low-rank representations of the right hand sides yield relatively large factors $G_{k,i}$ for the several stages. That is, the methods may cause unfeasible computational effort. See also Tables 6.6 and 6.8 for conviction. Still, there is a high potential for the application of the symmetric indefinite factorization.

Low-rank symmetric indefinite implicit Peer scheme for GDREs Consider the GALE (6.24), arising from the application of the implicit peer scheme together with Newton's method to the GDRE (6.1). Moreover, recall the corresponding right hand side

$$-\left(\tau_k g_{i,i} C_{k,i}^T C_{k,i} + \sum_{j=1}^{s} b_{i,j} E^T X_{k-1,j} E + \tau_k \sum_{j=1}^{s} a_{i,j} \mathcal{R}(t_{k-1,j}, X_{k-1,j}) + \tau_k \sum_{j=1}^{i-1} g_{i,j} \mathcal{R}(t_{k,j}, X_{k,j})\right)$$
$$- \tau_k g_{i,i} E^T X_{k,i}^{(\ell-1)} B_{k,i} B_{k,i}^T X_{k,i}^{(\ell-1)} E$$

for the peer stage i, within Newton step ℓ at time step k. In order to find a symmetric indefinite decomposition $-G_{k,i}^{(\ell)} S_{k,i}^{(\ell)} G_{k,i}^{(\ell)^T}$, we first define the factorizations for the Riccati operators. Similar to (6.64), introduced for the one-step Rosenbrock methods and with $X_{k,j} = L_{k,j} D_{k,j} L_{k,j}^T$, we obtain

$$\begin{aligned}
\mathcal{R}(t_{k,j}, X_{k,j}) &= \mathcal{T}_{k,j} \mathcal{M}_{k,j} \mathcal{T}_{k,j}^T, \\
\mathcal{T}_{k,j} &= \begin{bmatrix} C_{k,j}^T, & A_{k,j}^T L_{k,j}, & E^T L_{k,j} \end{bmatrix} \in \mathbb{R}^{n \times (q+2n_{L_{k,j}})}, \\
\mathcal{M}_{k,j} &= \begin{bmatrix} I_q & & \\ & & D_{k,j} \\ & D_{k,j} & -D_{k,j} L_{k,j}^T B_{k,j} B_{k,j}^T L_{k,j} D_{k,j} \end{bmatrix} \in \mathbb{R}^{(q+2n_{L_{k,j}}) \times (q+2n_{L_{k,j}})}.
\end{aligned}$$

Thus, the right hand side factors can be written in the form

$$G_{k,i}^{(\ell)} = \left[C_{k,i}^T,\ \ E^T L_{k-1,1}, \ldots, E^T L_{k-1,s},\ \ \mathcal{T}_{k-1,1}, \ldots, \mathcal{T}_{k-1,s},\ \ \mathcal{T}_{k,1}, \ldots, \mathcal{T}_{k,i-1},\ \ E^T X_{k,i}^{(\ell-1)} B_{k,i}\right],$$

$$S_{k,i}^{(\ell)} = \begin{bmatrix} \tau_k g_{i,i} I_q \\ & b_{i,1} D_{k-1,1} \\ && \ddots \\ &&& b_{i,s} D_{k-1,s} \\ &&&& \tau_k a_{i,1} \mathcal{M}_{k-1,1} \\ &&&&& \ddots \\ &&&&&& \tau_k a_{i,s} \mathcal{M}_{k-1,s} \\ &&&&&&& \ddots \\ &&&&&&&& \ddots \\ &&&&&&&&& \tau_k g_{i,1} \mathcal{M}_{k,1} \\ &&&&&&&&&& \ddots \\ &&&&&&&&&&& \tau_k g_{i,i-1} \mathcal{M}_{k,i-1} \\ &&&&&&&&&&&& \tau_k g_{i,i} I_m \end{bmatrix},$$

where $G_{k,i}^{(\ell)}$ is of column size

$$\begin{aligned} & q + \sum_{j=1}^{s} n_{L_{k-1,j}} + \sum_{j=1}^{s} (q + 2n_{L_{k-1,j}}) + \sum_{j=1}^{i-1} (q + 2n_{L_{k,j}}) + n_{L_{k,i}^{(\ell-1)}} \\ &= (s+i)q + 3\sum_{j=1}^{s} n_{L_{k-1,j}} + 2\sum_{j=1}^{i-1} n_{L_{k,j}} + m. \end{aligned}$$

Compared to (6.27) and assuming the column sizes of the classical low-rank solution factors and those of the symmetric indefinite scheme to be equal, this yields a saving of

$$(s+i-1)m + \sum_{j=1}^{s} n_{L_{k-1,j}} + \sum_{j=1}^{i-1} n_{L_{k,j}}$$

right hand side columns. Then, in the autonomous case and the GALE (6.28), we define the factors

$$\begin{aligned} \mathcal{T}_{k,j} &= \left[A^T L_{k,j},\ \ E^T L_{k,j}\right] \in \mathbb{R}^{n \times 2n_{L_{k,j}}}, \\ \mathcal{M}_{k,j} &= \begin{bmatrix} & D_{k,j} \\ D_{k,j} & -D_{k,j} L_{k,j}^T B_{k,j} B_{k,j}^T L_{k,j} D_{k,j} \end{bmatrix} \in \mathbb{R}^{2n_{L_{k,j}} \times 2n_{L_{k,j}}} \end{aligned}$$

and obtain the decomposition

$$G^{(\ell)}_{k,i} = \left[C^T,\ \ E^T L_{k-1,1},\dots,E^T L_{k-1,s},\ \ \mathcal{T}_{k-1,1},\dots,\mathcal{T}_{k-1,s},\ \ \mathcal{T}_{k,1},\dots,\mathcal{T}_{k,i-1},\ \ E^T X^{(\ell-1)}_{k,i} B\right],$$

$$S^{(\ell)}_{k,i} = \begin{bmatrix}
\tau_k(\sum_{j=1}^{s} a_{i,j}+\sum_{j=1}^{i} g_{i,j})I_q &&&&&&&&&&& \\
& b_{i,1}D_{k-1,1} &&&&&&&&&& \\
&& \ddots &&&&&&&&& \\
&&& b_{i,s}D_{k-1,s} &&&&&&&& \\
&&&& \tau_k a_{i,1}\mathcal{M}_{k-1,1} &&&&&&& \\
&&&&& \ddots &&&&&& \\
&&&&&& \tau_k a_{i,s}\mathcal{M}_{k-1,s} &&&&& \\
&&&&&&& \ddots &&&& \\
&&&&&&&& \tau_k g_{i,1}\mathcal{M}_{k,1} &&& \\
&&&&&&&&& \ddots && \\
&&&&&&&&&& \tau_k g_{i,i-1}\mathcal{M}_{k,i-1} & \\
&&&&&&&&&&& \tau_k g_{i,i} I_m
\end{bmatrix}$$

with $G^{(\ell)}_{k,i}$ being of column size

$$q + 3\sum_{j=1}^{s} n_{L_{k-1,j}} + 2\sum_{j=1}^{i-1} n_{L_{k,j}} + m.$$

Low-rank symmetric indefinite implicit Peer scheme for GDLEs In complete analogy, for the implicit peer method applied to a non-autonomous GDLE (6.2), we find the expressions

$$\begin{aligned}
\mathcal{L}(t_{k,j}, X_{k,j}) &= \mathcal{T}_{k,j}\mathcal{M}_{k,j}\mathcal{T}_{k,j}^T, \\
\mathcal{T}_{k,j} &= \begin{bmatrix} B_{k,j}, & A_{k,j}L_{k,j}, & EL_{k,j} \end{bmatrix} \in \mathbb{R}^{n\times(m+2n_{L_{k,j}})}, \\
\mathcal{M}_{k,j} &= \begin{bmatrix} I_m && \\ && D_{k,j} \\ & D_{k,j} & \end{bmatrix} \in \mathbb{R}^{(m+2n_{L_{k,j}})\times(m+2n_{L_{k,j}})}
\end{aligned}$$

for the Lyapunov operators. Consequently, the symmetric indefinite right hand side factors of the associated GALE (6.29) are given by

$$G_{k,i}^{(\ell)} = \left[B_{k,i}, \quad EL_{k-1,1}, \ldots, EL_{k-1,s}, \quad \mathcal{T}_{k-1,1}, \ldots, \mathcal{T}_{k-1,s}, \quad \mathcal{T}_{k,1}, \ldots, \mathcal{T}_{k,i-1}\right],$$

$$S_{k,i}^{(\ell)} = \begin{bmatrix} \tau_k g_{i,i} I_m & & & & & & & & & & \\ & b_{i,1} D_{k-1,1} & & & & & & & & & \\ & & \ddots & & & & & & & & \\ & & & b_{i,s} D_{k-1,s} & & & & & & & \\ & & & & \tau_k a_{i,1} \mathcal{M}_{k-1,1} & & & & & & \\ & & & & & \ddots & & & & & \\ & & & & & & \tau_k a_{i,s} \mathcal{M}_{k-1,s} & & & & \\ & & & & & & & \ddots & & & \\ & & & & & & & & \tau_k g_{i,1} \mathcal{M}_{k,1} & & \\ & & & & & & & & & \ddots & \\ & & & & & & & & & & \tau_k g_{i,i-1} \mathcal{M}_{k,i-1} \end{bmatrix}$$

with a total number of

$$(s+i)m + 3\sum_{j=1}^{s} n_{L_{k-1,j}} + 2\sum_{j=1}^{i-1} n_{L_{k,j}}$$

columns. Finally, for the autonomous DLE case, we have

$$G_{k,i}^{(\ell)} = \left[B, \quad EL_{k-1,1}, \ldots, EL_{k-1,s}, \quad \mathcal{T}_{k-1,1}, \ldots, \mathcal{T}_{k-1,s}, \quad \mathcal{T}_{k,1}, \ldots, \mathcal{T}_{k,i-1}\right],$$

$$S_{k,i}^{(\ell)} = \begin{bmatrix} \tau_k (\sum_{j=1}^{s} a_{i,j} + \sum_{j=1}^{i} g_{i,j}) I_m & & & & & & & & & \\ & b_{i,1} D_{k-1,1} & & & & & & & & \\ & & \ddots & & & & & & & \\ & & & b_{i,s} D_{k-1,s} & & & & & & \\ & & & & \ddots & & & & & \\ & & & & & \tau_k a_{i,1} \mathcal{M}_{k-1,1} & & & & \\ & & & & & & \ddots & & & \\ & & & & & & & \tau_k a_{i,s} \mathcal{M}_{k-1,s} & & \\ & & & & & & & & \tau_k g_{i,1} \mathcal{M}_{k,1} & \\ & & & & & & & & & \ddots \\ & & & & & & & & & & \tau_k g_{i,i-1} \mathcal{M}_{k,i-1} \end{bmatrix}$$

with the matrices

$$\mathcal{L}(t_{k,j}, X_{k,j}) = \mathcal{T}_{k,j}\mathcal{M}_{k,j}\mathcal{T}_{k,j}^T,$$

$$\mathcal{T}_{k,j} = \begin{bmatrix} AL_{k,j}, & EL_{k,j} \end{bmatrix} \in \mathbb{R}^{n\times 2n_{L_{k,j}}}, \quad \mathcal{M}_{k,j} = \begin{bmatrix} I_m & & \\ & & D_{k,j} \\ & D_{k,j} & \end{bmatrix} \in \mathbb{R}^{2n_{L_{k,j}}\times 2n_{L_{k,j}}}.$$

Hence, $G_{k,i}$ is of column size

$$m + 3\sum_{j=1}^{s} n_{L_{k-1,j}} + 2\sum_{j=1}^{i-1} n_{L_{k,j}}.$$

Low-rank symmetric indefinite Rosenbrock-type Peer scheme for GDREs For the low-rank symmetric indefinite factorization based solution of a non-autonomous GDRE (6.1), using the Rosenbrock-type peer method, we consider the GALE (6.32). Further, recall the associated simplified right hand side

$$\begin{aligned}\tilde{W}_{k,i} = {} & \tau_k \sum_{j=1}^{i-1} g_{i,j}\left(A_k^T X_{k,j} E + E^T X_{k,j} A_k\right) + E^T X_k B_k K_{k,i}^T + K_{k,i} B_k^T X_k E \\ & + \sum_{j=1}^{s-1}\Big(\tau_k a_{i,j}\big(C_{k-1,j}^T C_{k-1,j} - E^T X_{k-1,j} B_{k-1,j} B_{k-1,j}^T X_{k-1,j} E\big) \\ & + \check{A}_{k,i,j}^T X_{k-1,j} E + E^T X_{k-1,j}\check{A}_{k,i,j}\Big) + \tau_k a_{i,s} C_k^T C_k + b_{i,s} E^T X_k E,\end{aligned}$$

given in (6.33). Then, in analogy to (6.34), for the symmetric indefinite decomposition, we define

$$\begin{aligned}\mathcal{T}_{k,j} &= \begin{bmatrix} A_k^T L_{k,j}, & E^T L_{k,j} \end{bmatrix} \in \mathbb{R}^{n\times 2n_{L_{k,j}}},\ \mathcal{M}_{k,j} = \tau_k g_{i,j}\begin{bmatrix} 0 & D_{k,j} \\ D_{k,j} & 0 \end{bmatrix} \in \mathbb{R}^{2n_{L_{k,j}}\times 2n_{L_{k,j}}}, \\ \check{\mathcal{T}}_{k,i,j} &= \begin{bmatrix} C_{k-1,j}^T, & \check{A}_{k,i,j}^T L_{k-1,j}, & E^T L_{k-1,j} \end{bmatrix} \in \mathbb{R}^{n\times(q+2n_{L_{k-1,j}})}, \\ \check{\mathcal{M}}_{k,i,j} &= \begin{bmatrix} \tau_k a_{i,j} I_q & & \\ & & D_{k-1,j} \\ & D_{k-1,j} & -\tau_k a_{i,j} D_{k-1,j} L_{k-1,j}^T B_{k-1,j} B_{k-1,j}^T L_{k-1,j} D_{k-1,j} \end{bmatrix} \in \mathbb{R}^{(q+2n_{L_{k-1,j}})\times(q+2n_{L_{k-1,j}})}.\end{aligned}$$

Thus, the factors $G_{k,i}$ and $S_{k,i}$ read

$$G_{k,i} = \left[\mathcal{T}_{k,1}, \ldots, \mathcal{T}_{k,i-1},\ \ E^T X_k B_k,\ \ K_{k,i},\ \ \check{\mathcal{T}}_{k,i,1}, \ldots, \check{\mathcal{T}}_{k,i,s-1},\ \ C_k^T,\ \ E^T L_k\right],$$

$$S_{k,i} = \begin{bmatrix} \mathcal{M}_{k,1} & & & & & & & & & \\ & \ddots & & & & & & & & \\ & & \mathcal{M}_{k,i-1} & & & & & & & \\ & & & & I_m & & & & & \\ & & & I_m & & & & & & \\ & & & & & \mathcal{M}_{k,i,1} & & & & \\ & & & & & & \ddots & & & \\ & & & & & & & \mathcal{M}_{k,i,s-1} & & \\ & & & & & & & & \tau_k a_{i,s} I_q & \\ & & & & & & & & & b_{i,s} D_k \end{bmatrix}$$

with $G_{k,i}$ being of column size

$$\begin{aligned} &\sum_{j=1}^{i-1} 2n_{L_{k,j}} + 2m + \sum_{j=1}^{s-1} (q + 2n_{L_{k-1,j}}) + q + n_{L_k} \\ &= 2\sum_{j=1}^{i-1} n_{L_{k,j}} + 2\sum_{j=1}^{s-1} n_{L_{k-1,j}} + n_{L_k} + sq + 2m. \end{aligned}$$

For autonomous GDREs, the associated simplified right hand side (6.36), arising within the Rosenbrock-type peer scheme, yields the symmetric indefinite factors

$$G_{k,i} = \left[\mathcal{T}_{k,1}, \ldots, \mathcal{T}_{k,i-1},\ \ E^T X_k B,\ \ K_{k,i},\ \ C^T,\ \ E^T L_{k-1,1}, \ldots, E^T L_{k-1,s-1},\ \ E^T L_k\right],$$

$$S_{k,i} = \begin{bmatrix} \mathcal{M}_{k,1} & & & & & & & & & \\ & \ddots & & & & & & & & \\ & & \mathcal{M}_{k,i-1} & & & & & & & \\ & & & & I_m & & & & & \\ & & & I_m & & & & & & \\ & & & & & \tau_k \sum_{j=1}^{s} a_{i,j} I_q & & & & \\ & & & & & & \tilde{D}_{k-1,1} & & & \\ & & & & & & & \ddots & & \\ & & & & & & & & \tilde{D}_{k-1,s-1} & \\ & & & & & & & & & b_{i,s} D_k \end{bmatrix}$$

with the additionally defined matrices

$$\mathcal{T}_{k,j} = \left[A_k^T L_{k,j},\ \ E^T L_{k,j}\right] \in \mathbb{R}^{n \times 2n_{L_{k,j}}},\ \mathcal{M}_{k,j} = \tau_k g_{i,j} \begin{bmatrix} 0 & D_{k,j} \\ D_{k,j} & 0 \end{bmatrix} \in \mathbb{R}^{2n_{L_{k,j}} \times 2n_{L_{k,j}}},$$

$$\tilde{D}_{k-1,j} = b_{i,j} D_{k-1,j} - \tau_k a_{i,j} D_{k-1,j} L_{k-1,j} B B^T L_{k-1,j} D_{k-1,j}.$$

Here, the column size of the factor $G_{k,i}$ is

$$\sum_{j=1}^{i-1} 2n_{L_{k,j}} + 2m + q + \sum_{j=1}^{s} n_{L_{k-1,j}} = 2\sum_{j=1}^{i-1} n_{L_{k,j}} + \sum_{j=1}^{s} n_{L_{k-1,j}} + q + 2m.$$

For the modified Rosenbrock-type peer scheme applied to the non-autonomous GDRE, we consider the GALE (6.39) with the simplified right hand side (6.41). Assuming the auxiliary variables to admit an LRSIF decomposition $Y_{k,j} = \hat{L}_{k-1,j}\hat{D}_{k,j}\hat{L}_{k,j}$, the associated symmetric indefinite formulation is given by the factors

$$\begin{aligned}
G_{k,i} = \Big[&\check{\mathfrak{I}}_{k-1,i,1}, \dots, \check{\mathfrak{I}}_{k-1,i,s},\ E^T X_k B_k,\ K_{k,i},\ \mathfrak{T}_{k-1,1}, \dots, \mathfrak{T}_{k-1,s-1}, \\
&C_k^T, A_k^T L_k,\ E^T L_k, E^T\hat{L}_{k,1}, \dots, E^T\hat{L}_{k,i-1},\Big],
\end{aligned}$$

$$S_{k,i} = \begin{bmatrix}
-\check{\mathcal{M}}_{k-1,i,1} &&&&&&&&&&&&& \\
& \ddots &&&&&&&&&&&& \\
&& -\check{\mathcal{M}}_{k-1,i,s} &&&&&&&&&&& \\
&&&& I_m &&&&&&&&& \\
&&& I_m &&&&&&&&&& \\
&&&&& a_{i,1}\mathcal{M}_{k-1,1} &&&&&&&& \\
&&&&&& \ddots &&&&&&& \\
&&&&&&& a_{i,s-1}\mathcal{M}_{k-1,s-1} &&&&&& \\
&&&&&&&& \ddots &&&&& \\
&&&&&&&&& a_{i,s}I_q &&&& \\
&&&&&&&&&&& a_{i,s}D_k && \\
&&&&&&&&&& a_{i,s}D_k &&& \\
&&&&&&&&&&&& -\frac{g_{i,1}}{\tau_k}\hat{D}_{k,1} & \\
&&&&&&&&&&&&& \ddots \\
&&&&&&&&&&&&&& -\frac{g_{i,i-1}}{\tau_k}\hat{D}_{k,i-1}
\end{bmatrix}$$

with

$$\begin{aligned}
\check{\mathfrak{I}}_{k-1,i,j} &= \left[\check{A}_{k,i,j}^T\hat{L}_{k-1,j},\ \ E^T\hat{L}_{k-1,j}\right] \in \mathbb{R}^{n\times 2n_{L_{k-1,j}}}, \quad \check{\mathcal{M}}_{k-1,i,j} = \begin{bmatrix} & \hat{D}_{k-1,j} \\ \hat{D}_{k-1,j} & \end{bmatrix} \in \mathbb{R}^{2n_{L_{k-1,j}}\times 2n_{L_{k-1,j}}}, \\
\mathfrak{T}_{k-1,j} &= \left[C_{k-1,j}^T,\ \ A_{k-1,j}^T L_{k-1,j},\ \ E^T L_{k-1,j}\right] \in \mathbb{R}^{n\times(q+2n_{L_{k-1,j}})}, \\
\mathcal{M}_{k-1,j} &= \begin{bmatrix} I_q && \\ && D_{k-1,j} \\ & D_{k-1,j} & -D_{k-1,j}L_{k-1,j}^T B_{k-1,j}B_{k-1,j}^T L_{k-1,j}D_{k-1,j} \end{bmatrix} \in \mathbb{R}^{(q+2n_{L_{k-1,j}})\times(q+2n_{L_{k-1,j}})},
\end{aligned}$$

defining the factorization of the Lyapunov-type expression and the Riccati operators, respectively. The resulting column size of $G_{k,i}$ is then given by

$$\begin{aligned}
&\sum_{j=1}^{s} 2n_{\hat{L}_{k-1,j}} + 2m + \sum_{j=1}^{s-1}(q + 2n_{L_{k-1,j}}) + q + 2n_{L_k} + \sum_{j=1}^{i-1} n_{\hat{L}_{k,j}} \\
&= \sum_{j=1}^{i-1} n_{\hat{L}_{k,j}} + 2\sum_{j=1}^{s}(n_{\hat{L}_{k-1,j}} + n_{L_{k-1,j}}) + sq + 2m.
\end{aligned}$$

For the autonomous case and the associated GALE (6.43) and its condensed right hand side (6.47), we find the factors

$$\begin{aligned}
G_{k,i} = \Big[&C^T,\ E^T X_{k-1,1}B, \dots, E^T X_{k-1,s-1}B,\ E^T X_k B,\ K_{k,i}, \\
&E^T\hat{L}_{k-1,1}, \dots, E^T\hat{L}_{k-1,s},\ E^T\hat{L}_{k,1}, \dots, E^T\hat{L}_{k,i-1}\Big],
\end{aligned}$$

$$S_{k,i} = \begin{bmatrix}
\sum_{j=1}^{s} a_{i,j} I_q & & & & & & & & & & & & \\
& -a_{i,1} I_m & & & & & & & & & & & \\
& & \ddots & & & & & & & & & & \\
& & & -a_{i,s-1} I_m & & & & & & & & & \\
& & & & & I_m & & & & & & & \\
& & & & I_m & & & & & & & & \\
& & & & & & \frac{\mathbf{b}_{i,1}}{\tau_k}\hat{D}_{k-1,1} & & & & & & \\
& & & & & & & \ddots & & & & & \\
& & & & & & & & \frac{\mathbf{b}_{i,s}}{\tau_k}\hat{D}_{k-1,s} & & & & \\
& & & & & & & & & \ddots & & & \\
& & & & & & & & & & -\frac{\mathbf{g}_{i,1}}{\tau_k}\hat{D}_{k,1} & & \\
& & & & & & & & & & & \ddots & \\
& & & & & & & & & & & & -\frac{\mathbf{g}_{i,i-1}}{\tau_k}\hat{D}_{k,i-1}
\end{bmatrix},$$

where $G_{k,i}$ is of column size

$$q + \sum_{j=1}^{s-1} m + 2m + \sum_{j=1}^{s} n_{\hat{L}_{k-1,j}} + \sum_{j=1}^{i-1} n_{\hat{L}_{k,j}} = \sum_{j=1}^{i-1} n_{\hat{L}_{k,j}} + \sum_{j=1}^{s} n_{\hat{L}_{k-1,j}} + q + (s+1)m.$$

Compared to the classical low-rank representation, we have a saving of "only" m right hand side columns. Still, the factors of the classical scheme are complex which is avoided by the symmetric indefinite formulation.

Low-rank symmetric indefinite Rosenbrock-type Peer scheme for GDLEs In the non-autonomous DLE case, we again consider the GALE (6.50). Defining the auxiliary factors

$$\mathcal{T}_{k,j} = \begin{bmatrix} A_k L_{k,j}, & E L_{k,j} \end{bmatrix}, \qquad \mathcal{M}_{k,j} = \tau_k g_{i,j} \begin{bmatrix} & D_{k,j} \\ D_{k,j} & \end{bmatrix},$$

$$\check{\mathcal{T}}_{k,i,j} = \begin{bmatrix} B_{k-1,j} & \check{A}_{k,i,j} L_{k-1,j}, & E Z_{k-1,j} \end{bmatrix}, \qquad \check{\mathcal{M}}_{k,i,j} = \begin{bmatrix} \tau_k a_{i,j} I_m & & \\ & & D_{k-1,j} \\ & D_{k-1,j} & \end{bmatrix}$$

with $\mathcal{T}_{k,j} \in \mathbb{R}^{n \times 2n_{L_{k,j}}}$, $\mathcal{M}_{k,j} \in \mathbb{R}^{2n_{L_{k,j}} \times 2n_{L_{k,j}}}$ and $\check{\mathcal{T}}_{k,i,j} \in \mathbb{R}^{n \times (m+2n_{L_{k-1,j}})}$, $\check{\mathcal{M}}_{k,i,j} \in \mathbb{R}^{(m+2n_{L_{k-1,j}}) \times (m+2n_{L_{k-1,j}})}$ for the Lyapunov-type expressions in the right hand side, we end up with a symmetric indefinite decomposition of the form

$$G_{k,i} = \begin{bmatrix} \mathcal{T}_{k,1}, \ldots, \mathcal{T}_{k,i-1}, & B_k, & E L_k, & \check{\mathcal{T}}_{k,i,1}, \ldots, \check{\mathcal{T}}_{k,i,s-1} \end{bmatrix},$$

$$S_{k,i} = \begin{bmatrix} \mathcal{M}_{k,1} & & & & & & & \\ & \ddots & & & & & & \\ & & \mathcal{M}_{k,i-1} & & & & & \\ & & & \tau_k a_{i,s} I_m & & & & \\ & & & & b_{i,s} D_k & & & \\ & & & & & \check{\mathcal{M}}_{k,i,1} & & \\ & & & & & & \ddots & \\ & & & & & & & \check{\mathcal{M}}_{k,i,s-1} \end{bmatrix},$$

where the column size of $G_{k,i}$ is given by

$$2\sum_{j=1}^{i-1} n_{L_{k,j}} + 2\sum_{j=1}^{s-1} n_{L_{k-1,j}} + n_{L_{k-1,s}} + sm.$$

Considering the GALE (6.52) arising within the scheme applied to autonomous GDLEs, the symmetric indefinite factors read

$$G_{k,i} = \begin{bmatrix} \mathcal{T}_{k,1}, \ldots, \mathcal{T}_{k,s}, & B, & E L_{k-1,1}, \ldots, E L_{k-1,s} \end{bmatrix},$$

$$S_{k,i} = \begin{bmatrix} \tau_k g_{i,1} \mathcal{M}_{k,1} & & & & & & \\ & \ddots & & & & & \\ & & \tau_k g_{i,i-1} \mathcal{M}_{k,i-1} & & & & \\ & & & \tau_k \sum_{j=1}^{s} a_{i,j} I_m & & & \\ & & & & b_{i,1} D_{k-1,1} & & \\ & & & & & \ddots & \\ & & & & & & b_{i,s} D_{k-1,s} \end{bmatrix}$$

with

$$\mathcal{T}_{k,j} = \begin{bmatrix} A L_{k,j}, & E L_{k,j} \end{bmatrix}, \qquad \mathcal{M}_{k,j} = \tau_k g_{i,j} \begin{bmatrix} & D_{k,j} \\ D_{k,j} & \end{bmatrix}$$

and a column size of

$$2\sum_{j=1}^{i-1} n_{L_{k,j}} + \sum_{j=1}^{s} n_{L_{k-1,j}} + m.$$

For the modified Rosenbrock-type peer scheme applied to non-autonomous GDLEs and the associated GALE (6.54), we define the matrices

$$\mathcal{T}_{k-1,j} = \begin{bmatrix} B_{k-1,j}, & A_{k-1,j}L_{k-1,j}, & EL_{k-1,j} \end{bmatrix}, \qquad \mathcal{M}_{k-1,j} = \tau_k a_{i,j} \begin{bmatrix} I_m & & \\ & & D_{k-1,j} \\ & D_{k-1,j} & \end{bmatrix},$$

$$\check{\mathcal{T}}_{k-1,i,j} = \begin{bmatrix} \check{A}_{k,i,j}\hat{L}_{k-1,j}, & E\hat{L}_{k-1,j}, \end{bmatrix}, \qquad \check{\mathcal{M}}_{k-1,i,j} = -\begin{bmatrix} & \hat{D}_{k-1,j} \\ \hat{D}_{k-1,j} & \end{bmatrix}$$

and therefore obtain the factorization

$$G_{k,i} = \begin{bmatrix} \check{\mathcal{T}}_{k,i,1}, \dots, \check{\mathcal{T}}_{k,i,s}, & \mathcal{T}_{k,1}, \dots, \mathcal{T}_{k,s}, & E\hat{L}_{k,1}, \dots, E\hat{L}_{k,i-1} \end{bmatrix},$$

$$S_{k,i} = \begin{bmatrix} \check{\mathcal{M}}_{k,i,1} & & & & & & & & \\ & \ddots & & & & & & & \\ & & \check{\mathcal{M}}_{k,i,s} & & & & & & \\ & & & \mathcal{M}_{k,1} & & & & & \\ & & & & \ddots & & & & \\ & & & & & \mathcal{M}_{k,s} & & & \\ & & & & & & -\frac{g_{i,1}}{\tau_k}\hat{D}_{k,1} & & \\ & & & & & & & \ddots & \\ & & & & & & & & -\frac{g_{i,i-1}}{\tau_k}\hat{D}_{k,i-1} \end{bmatrix},$$

for the right hand side $\tilde{W}_{k,i}$. The column size of $G_{k,i}$ is given by

$$\sum_{j=1}^{i-1} n_{\hat{L}_{k,j}} + 2\sum_{j=1}^{s} \left(n_{\hat{L}_{k-1,j}} + n_{L_{k-1,j}} \right) + sm.$$

Finally, for the GALE (6.55), arising within the modified Rosenbrock-type peer scheme applied to an autonomous GDLE, the symmetric indefinite factorization is given in the form

$$G_{k,i} = \left[B, \quad E\hat{Z}_{k-1,1}, \ldots, E\hat{Z}_{k-1,s} E\hat{Z}_{k,1}, \ldots, E\hat{Z}_{k,i-1},\right],$$

$$S_{k,i} = \begin{bmatrix} \sum_{j=1}^{s} a_{i,j} I_m & & & & & & \\ & \frac{\mathbf{b}_{i,1}}{\tau_k} \hat{D}_{k-1,1} & & & & & \\ & & \ddots & & & & \\ & & & \frac{\mathbf{b}_{i,s}}{\tau_k} \hat{D}_{k-1,s} & & & \\ & & & & -\frac{\mathbf{g}_{i,1}}{\tau_k} \hat{D}_{k,1} & & \\ & & & & & \ddots & \\ & & & & & & -\frac{\mathbf{g}_{i,i-1}}{\tau_k} \hat{D}_{k,i-1} \end{bmatrix}$$

with $G_{k,i}$ being of column size

$$m + \sum_{j=1}^{s} n_{\hat{L}_{k-1,j}} + \sum_{j=1}^{i-1} n_{\hat{L}_{k,j}}.$$

Here, we note that in comparison to the classical factorization, no re-arrangement and therefore no significant reduction of the column blocks is achieved. Still, the use of complex data and thus the need for complex arithmetic and storage can be avoided.

6.3. Column Compression

Having a deeper look at the (G)ALE solution methods, presented in Section 2.1.4, one easily observes that the size of the solution factors Z_k, L_k at time t_k is proportional to the size of the associated right hand side factor G_k of the (G)ALE. Moreover, from Sections 6.1 and 6.2, we know that in the subsequent step of the time integration method the factors Z_k, L_k enter the right hand side factor G_{k+1} at time t_{k+1} and so on. That is, the number of columns of the solution and the right hand side factors cyclically depend on each other. Therefore, the sizes of both, the solution factor, as well as the right hand side factor will, driven by the solution method, increase drastically over time. Still, the solution $X(t)$ of the DMEs is expected to be of low numerical rank for all $t \in [t_0, t_f]$. That is, the number of linear independent columns of the solution factors Z_k, L_k is limited by that rank. Therefore, if the column size of Z_k, L_k exceeds the actual rank of the solution, an elimination of redundant information in terms of a column compression based on the numerical rank of the factors becomes necessary. Note that, a priori, the rank of the solution is unknown and a compression criterion is to be defined. However, for the presented integration methods, it appeared to be useful to apply a column compression right after setting up every right hand side factor and after every computation of the solution factors Z_k or L_k.

In the following, we investigate three different types of column compression techniques. First, we consider the rank-revealing QR decomposition and SVD based column compression approaches for the classical low-rank factorizations. Second, an adaptation of these well-known compression procedures to complex factors, defining a real symmetric product and finally, an approach dealing with the symmetric indefinite decomposition, are presented. These approaches are employed in MATLAB notation, in order to specify sub blocks of a matrix. Note that the rank r in practice needs to be decided numerically. Still, the results are presented for exact computations.

6.3.1. Classical Low-Rank Compression

Given a matrix $G \in \mathbb{R}^{n\times k}$, defining a real symmetric product $GG^T \in \mathbb{R}^{n\times n}$. Then, the numerical rank $r \leq k$ can be computed, using the following approaches.

QR based column compression

i) Compute $G^T = QR\Pi^T$ with $G \in \mathbb{R}^{n\times k}$, $Q \in \mathbb{R}^{k\times k}$, $Q^TQ = I_k$, $R \in \mathbb{R}^{k\times n}$ and a permutation matrix $\Pi \in \mathbb{R}^{n\times n}$.

ii) Set $G_r = \Pi R_r^T \in \mathbb{R}^{n\times r}$, where $r = \text{rank}(R)$ and $R_r = R(1:r,:) \in \mathbb{R}^{r\times n}$, $Q_r = Q(1:r,1:r) \in \mathbb{R}^{r\times r}$, such that

$$G_rG_r^T = \Pi R_r^T R_r \Pi^T = \Pi R_r^T Q_r^T Q_r R_r \Pi^T = \Pi R^T Q^T QR\Pi^T = GG^T.$$

SVD based column compression

i) Compute $G = U\Sigma V^T$ with $G \in \mathbb{R}^{n\times k}, U \in \mathbb{R}^{n\times k}$, $U^TU = I_k$, $\Sigma \in \mathbb{R}^{k\times k}$ and $V \in \mathbb{R}^{k\times k}$, $V^TV = I_k$.

ii) Set $G_r = U_r\Sigma_r \in \mathbb{R}^{n\times r}$, where $r = \text{rank}(R)$, $U_r = U(:,1:r) \in \mathbb{R}^{n\times r}$, $\Sigma_r = \Sigma(1:r,1:r) \in \mathbb{R}^{r\times r}$, and $V_r = V(:,1:r) \in \mathbb{R}^{k\times r}$, such that

$$G_rG_r^T = U_r\Sigma_r^2U_r^T = U_r\Sigma_rV_r^TV_r\Sigma_rU_r^T = U\Sigma V^TV\Sigma U^T = GG^T.$$

Motivated by the integration methods of order $s \geq 2$, now we consider a complex matrix $G \in \mathbb{C}^{n\times k}$, still defining a real symmetric product GG^T.

6.3.2. Classical Low-Rank Compression for Complex Data

Let $G \in \mathbb{C}^{n\times k}$. Therefore, the QR decomposition similar to the real case reads

$$G^T = QR\Pi^T \text{ with } Q \in \mathbb{C}^{k\times k},\ Q^HQ = I_k,\ R \in \mathbb{C}^{k\times n} \text{ and } \Pi \in \mathbb{R}^{n\times n}.$$

Since for complex data Q is unitary and therefore $Q^TQ \neq I_k$, we have

$$GG^T = \Pi R^T Q^T Q R \Pi^T \neq \Pi R_r^T R_r \Pi^T$$

and the compressed factor G_r cannot simply be defined in the form $G_r = \Pi R_r^T \in \mathbb{C}^{n\times r}$. A similar problem occurs in the case of the SVD based approach. Computing

$$G = U\Sigma V^H \text{ with } U \in \mathbb{C}^{n\times k},\ U^HU = I_k,\ \Sigma \in \mathbb{C}^{k\times k},\ V \in \mathbb{C}^{k\times k},\ V^HV = I_k$$

and following Step ii) from the SVD based compression for real data, we obtain

$$G_rG_r^T = U_r\Sigma_rV^H\bar{V}\Sigma_rU_r^T \neq U_r\Sigma_rV^HV\Sigma_rU_r^T = U_r\Sigma_r^2U_r^T.$$

Clearly, using the matrix $G \in \mathbb{C}^{n\times k}$ in the real symmetric and indefinite product $GG^T \in \mathbb{R}^{n\times n}$ requires us to properly adjust the compression to the outer product in use. We propose the following procedure:

i) Compute $G = QR\Pi^T$ with $Q \in \mathbb{C}^{n\times k}$, $Q^HQ = I_k$, $R \in \mathbb{C}^{k\times k}$, and the permutation matrix $\Pi \in \mathbb{R}^{k\times k}$.

ii) Compute a decomposition $R\Pi^T\Pi R^T = RR^T = V\Lambda V^T$ with $V \in \mathbb{C}^{k\times k}$, $V^TV = I_k$ and a diagonal matrix $\Lambda \in \mathbb{C}^{k\times k}$ with diagonal entries $|\lambda_1| \geq |\lambda_2| \geq \cdots \geq |\lambda_k|$.

iii) Set the compressed factor $G_r = QV_r\Lambda_r^{\frac{1}{2}} \in \mathbb{C}^{n\times r}$ with $r \leq k$ and $|\lambda_{r+1}| \leq \varepsilon$.

The remaining question is to find an orthogonal matrix V, such that the complex symmetric matrix $RR^T \in \mathbb{C}^{k\times k}$ is diagonalized. In fact, the decomposition required in Step ii) of the above procedure is a so-called *Takagi factorization*, see e.g., [104, Corollary 4.4.4]. Due to the absence of a Takagi factorization routine in the common linear algebra packages, here we propose the following approach:

Compute the eigendecomposition

$$RR^T = \tilde{V}\Lambda\tilde{V}^{-1}$$

of the complex symmetric matrix RR^T. Then, in any software tool based on LAPACK [5], the columns $\tilde{v}_i$ in $\tilde{V}$ are normalized but not necessarily orthogonal with respect to the complex inner product. If now $\tilde{v}_i \in \mathbb{C}^k\backslash\mathbb{R}^k$, the properties

$$\begin{aligned} \tilde{v}_i^*\tilde{v}_i &= 1 \quad \Rightarrow \quad \tilde{v}_i^T\tilde{v}_i \neq 1, \\ \tilde{v}_i^*\tilde{v}_j &\neq 0. \end{aligned} \tag{6.65}$$

are satisfied. From the principle of biorthogonality, see e.g., [104, Theorem 1.4.7 and Proof of Theorem 4.4.13], we know that the eigenvectors $\tilde{v}_i, \tilde{v}_j$ of the symmetric matrix RR^T are orthogonal with respect to the real inner product, i.e., $\tilde{v}_i^T\tilde{v}_j = 0$ for $i \neq j$. Since the right hand side is constructed to be of the form $GG^T = QV\Lambda V^T Q^T$, it remains to ensure $RR^T = V\Lambda V^T$. Using (6.65) and the orthogonality of $\tilde{v}_i, \tilde{v}_j$, we have

$$\tilde{V}^T \tilde{V} = \begin{bmatrix} \tilde{v}_1^T \tilde{v}_1 & & 0 \\ & \ddots & \\ 0 & & \tilde{v}_k^T \tilde{v}_k \end{bmatrix}.$$

That is, the eigenvectors $\tilde{v}_i$ need to be normalized with respect to the real inner product. Defining $V = \tilde{V}\tilde{D}$ with $\tilde{D} = \operatorname{diag}\left(\sqrt{\tilde{v}_1^T \tilde{v}_1}, \ldots, \sqrt{\tilde{v}_k^T \tilde{v}_k}\right)^{-1}$ yields,

$$\begin{aligned} V^T V &= \tilde{D}\tilde{V}^T \tilde{V}\tilde{D} = I_k \\ \Leftrightarrow \quad \tilde{D}\tilde{V}^T &= V^T = V^{-1} = \tilde{D}^{-1}\tilde{V}^{-1}. \end{aligned} \tag{6.66}$$

Then, for the eigendecomposition of RR^T, we have

$$RR^T = \tilde{V}\Lambda\tilde{V}^{-1} = \tilde{V}\Lambda\tilde{D}\tilde{D}^{-1}\tilde{V}^{-1}$$

and since the matrices Λ, $\tilde{D}$ are diagonal and thus commute ($\Lambda\tilde{D} = \tilde{D}\Lambda$), with (6.66), we further obtain

$$\begin{aligned} RR^T &= \tilde{V}\Lambda\tilde{V}^{-1} = \tilde{V}\Lambda\tilde{D}\tilde{D}^{-1}\tilde{V}^{-1} \\ &= \tilde{V}\tilde{D}\Lambda\tilde{D}^{-1}\tilde{V}^{-1} = V\Lambda V^{-1} = V\Lambda V^T. \end{aligned}$$

That is, scaling the eigenvectors $\tilde{v}_i$ with $(\tilde{v}_i^T \tilde{v}_i)^{-\frac{1}{2}}$, $i = 1, \ldots, k$, the eigenspace of the complex symmetric matrix RR^T is not changed and we end up with the desired Takagi factorization

$$RR^T = V\Lambda V^T.$$

Note that for normalized real eigenvectors $\tilde{v}_i \in \mathbb{R}^k$ we have $\tilde{v}_i^T \tilde{v}_i = 1$ and thus the scaling is redundant.

6.3.3. Symmetric Indefinite Compression

As for the classical low-rank integration methods, the symmetric indefinite low-rank factors of the right hand side and the solution will cyclically grow within every time integration step. That is, we also need to perform a column compression in order to reveal the numerical rank of the corresponding factor. Following the statements in [38, Section 6.3.3] the factors $G \in \mathbb{R}^{n \times k}$, $S = S^T \in \mathbb{R}^{k \times k}$, defining the real symmetric product GSG^T, can be compressed as follows:

i) Compute $G = QR\Pi^T$ with $Q \in \mathbb{R}^{n \times k}, R \in \mathbb{R}^{k \times k}$ and $\Pi \in \mathbb{R}^{k \times k}$.

ii) Compute a decomposition $R\Pi^T S \Pi R^T = V \Lambda V^T$ with $V \in \mathbb{R}^{k \times k}$ and a diagonal matrix $\Lambda \in \mathbb{R}^{k \times k}$ with diagonal entries $|\lambda_1| \geq |\lambda_2| \geq \cdots \geq |\lambda_k|$.

iii) Set the compressed factors $G_r = QV_r \in \mathbb{R}^{n \times r}$, $S_r = \Lambda_r$ with $r \leq k$ and $|\lambda_{r+1}| \leq \varepsilon$.

Since $R\Pi^T S \Pi R^T \in \mathbb{R}^{k \times k}$ is symmetric, a decomposition $V \Lambda V^T$ always exists and can e.g., be computed via an eigendecomposition. Thus, comparing the computational cost, the above procedure is equal to the classical low-rank column compression for complex data, provided that the sizes of the thin rectangular matrices coincide.

6.4. Numerical Experiments

In this section we investigate the solution of the GDRE

$$\begin{aligned} E^T \dot{X}(t) E &= -C^T Q C - A^T X(t) E - E^T X(t) A + E^T X(t) B R^{-1} B^T X(t) E, \\ E^T X(t_f) E &= C^T S C. \end{aligned} \tag{6.67}$$

In [149] the general behavior of most of the presented ODE solvers, based on dense solution methods applied to small-scale examples, is investigated. Therefore, here we mainly focus on the comparison of the LRCF and LRSIF factorization schemes.

The weighting matrices Q, R and S are chosen to be identities of appropriate size. Further, note that the GDRE arising from an LQR problem has to be solved backwards in time. Therefore, the following results need to be interpreted starting from the end point of the corresponding time interval. Moreover, due to the high memory requirements for storing the full dense solution $X(t)$, the evaluation of the results is restricted to one component $K_{i,j}(t)$ of the feedback matrix $K(t) = -B^T X(t) E$ given in Equation (3.23) with $i = 1, \ldots, m$, $j = 1, \ldots, n$. The selected components $K_{i,j}$ are chosen with respect to a relatively large amplitude of their trajectories such that differences are well visible. The relative errors, as well as the deviation of the two low-rank schemes in the remainder are given in the Frobenius norm $||.||_F$ for the full feedback K over the entire time interval. That is, the errors of the different low-rank schemes compared to a reference solution or the deviation of both low-rank representations are computed in the form

$$\frac{||K_{ref} - K_{LRCF/LRSIF}||_F}{||K_{ref}||_F} \quad \text{or} \quad \frac{||K_{LRCF} - K_{LRSIF}||_F}{||K_{LRCF}||_F}, \tag{6.68}$$

respectively. Since the advantage of the LRSIF Krylov subspace Lyapunov solver compared to its LRCF counterpart is rather obvious, see Algorithms 2.3 and 2.4, the following experiments, for the comparison of both low-rank strategies, are restricted to the DRE solvers using the ADI based Lyapunov solvers, given by the Algorithms 2.1 and 2.2, respectively. The results are entirely computed, using the following stopping criteria for Newton's method and the ADI iteration. A Newton tolerance of $1e$-10

Time integration method	Acronym
BDF of order s	BDF(s)
Rosenbrock of order s	Ros(s)
Midpoint rule	Mid
Trapezoidal rule	Trap
Implcit peer of order s	Peer(s)
Rosenbrock-type peer of order s	RosPeer(s)
Modified RosPeer(s)	mRosPeer(s)

Table 6.1.: Acronyms of the time integration methods ($s = 1, \dots, 4$).

for the relative feedback change in the sense of the Frobenius norms given in Equation (6.68) and a maximum number of 15 Newton steps are implemented. For the Lyapunov residual within the ADI iteration, we use a tolerance of $n\varepsilon$, where n is the problem dimension and ε denotes the machine precision. Further, a maximum number of 100 ADI steps is allowed. For the column compression techniques the same truncation tolerance of $n\varepsilon$ was used.

For better readability, Table 6.1 recapitulates the acronyms of the different integration methods that were introduced in the corresponding subsections of Chapter 5.

6.4.1. Autonomous Control Systems

LTI Rail Profile

The first example is again the rail example [34, 168], previously introduced in Section 4.3.3. Here, we consider the original LTI system matrices and the associated GDRE on the simulation time interval $[0,\ 4\,500]$ s. The solution to the GDRE is computed with a constant time step $\tau = 11.25$ s, resulting in 400 time steps. Recall that the time has to be scaled by $1e2$ such that a real time interval of $[0, 45]$ s and a step size τ of 0.1125 s are investigated. In order to be able to compare the results of the different low-rank GDRE solvers to a reference solution from a dense fourth-order Rosenbrock (Ros(4)) method [182], the smallest available discretization level with $n = 371$ is chosen. The Ros(4) procedure used here, is presented in the Appendix A. Moreover, the Ros(4) solution is computed by the *Parareal* based time integration scheme considered in [119].

Figures 6.1(a) and 6.1(b) present the evolution of the feedback component $K_{1,71}$ for the Ros(4) reference solution and all symmetric indefinite factorization based solvers considered above. The associated relative errors, with respect to the reference, are depicted in Figure 6.1(c). The peaks in the relative error regarding the reference solution at the end of the time interval and therefore the beginning of the solution horizon of the GDRE result from the zero initial condition and the trajectory evolving close to zero for the first 3 seconds. Figure 6.1(d) presents the relative deviation of the low-rank strategies.

A comparison of the computational times, the acceleration factor comparing both low-rank formulations, the relative errors with respect to the reference solution and the relative deviation, computed in the Frobenius norm, for the LRCF and LRSIF schemes are given in Table 6.2. The almost identical relative errors of the LRCF and LRSIF methods and in particular the speedup factor of at least 2 confirm the predicted dominance of the symmetric indefinite factorization based DRE solvers. It can also be seen that, in contrast to the theoretical assumption, the class of peer methods can compete with the standard ODE solution methods with respect to both, the computational times and the achieved relative errors. In particular, for this example, the LRSIF version of the mRosPeer(1) method yields best computational time and the RosPeer(2) and mRosPeer(2) generate the smallest relative error with respect to the reference solution. The highest speedup is achieved by the LRSIF Ros(2) scheme.

Exploiting the existence of the Ros(4) reference solution, in Figure 6.2 an accuracy plot (a), as well as an efficiency plot (b) for $25, 50, 100, 200$ and 400 steps, resulting in the step sizes $1.8, 0.9, 0.45, 0.225$ and 0.1125, respectively, are depicted. The efficiency plot reveals that for this example the symmetric indefinite versions of the RosPeer(2) and mRosPeer(2) integration methods show the best performance.

In addition to the computation times and in order to investigate the performance in more detail, Figure 6.3 shows the evolution of the number of Newton steps needed by the classical and symmetric indefinite low-rank solvers. It can be seen that, for this example, the number of Newton iterations is invariant with respect to the time integration methods and the chosen factorization strategy. In Figure 6.4 the total numbers of ADI steps over the entire time horizon are given. The horizontal separations have the following meaning. For the nonlinear integration schemes (BDF(1), BDF(2), Mid, Trap) the stacked bars correspond to the several Newton steps. For the Ros, RosPeer, and mRosPeer methods of order one and two, the different stages are indicated by the hatching schemes and the Peer(1) and Peer(2) approaches are marked by a mixture of both. We observe that the symmetric indefinite Lyapunov solvers save a number of ADI steps, except for the RosPeer(1) method and its modified version mRosPeer(1). In addition to the savings of the shifted linear system solves within the ADI, this directly explains the computation times in Table 6.2. Figure 6.5 shows the average number of ADI steps required per time step related to the step size used for the time integration process. It can be observed that decreasing the time step size, the number of ADI steps decreases as well. This is due to the fact, that the symmetric positive definite mass matrix E within the linear systems with coefficient matrices of the form

$$\tau_k \alpha F_k - \beta E, \qquad \alpha F_k - \frac{\beta}{\tau_k} E$$

that have to be solved at every ADI step dominates the spectrum of eigenvalues for small time step sizes τ. That is, the real parts of the eigenvalues of the linear systems move towards -∞ for decreasing time step sizes τ. The parameters α and β, as well as the matrix F_k are defined by the chosen time integration methods and can be found in the corresponding Sections 5.1- 5.4.

Last but not least, in Figure 6.6 the numerical ranks, based on the column compression tolerance $n\varepsilon$, for both low-rank strategies are compared to the numerical rank of the dense reference solution. Recall, that the GDRE (6.67) has to be solved backwards in time and by its final condition $X(t_f) = E^{-1}C^TSCE^{-1}$, the numerical rank of X at t_f is given by the rank of C. For the classical low-rank based integration schemes, given in Figure 6.6(a), a significant mismatch between the first-order solutions of the BDF(1), Ros(1) and Peer(1) schemes and the remaining methods can be observed. Excluding these three first-order integrators, the numerical ranks of the LRCF solution methods almost recover the rank evolution of the reference solution. Regarding the symmetric indefinite based ODE solvers, presented in Figure 6.6(b) no outliers are observed except for the Ros(1) methods. To be more precise, the LRSIF Ros(1) method shows a highly oscillatory behavior, still being more accurate then the LRCF BDF(1), Ros(1) and Peer(1) results. Comparing both low-rank strategies it can be observed that, except for the outliers, the results are qualitatively equal. In other words, the numerical ranks of the solutions, i.e., the size of the low-rank solution factors coincide with respect to most of the integration schemes. Therefore, the differences in the computational times can be traced back to the integration scheme dependent growth of the right hand side factors, summarized in Tables 6.6 and 6.7 for the DRE. That is, the computational cost for the application of the column compression techniques and the solution of the innermost Lyapunov equations directly depends on the size of the right hand side factors, entering the Lyapunov solver at every time step.

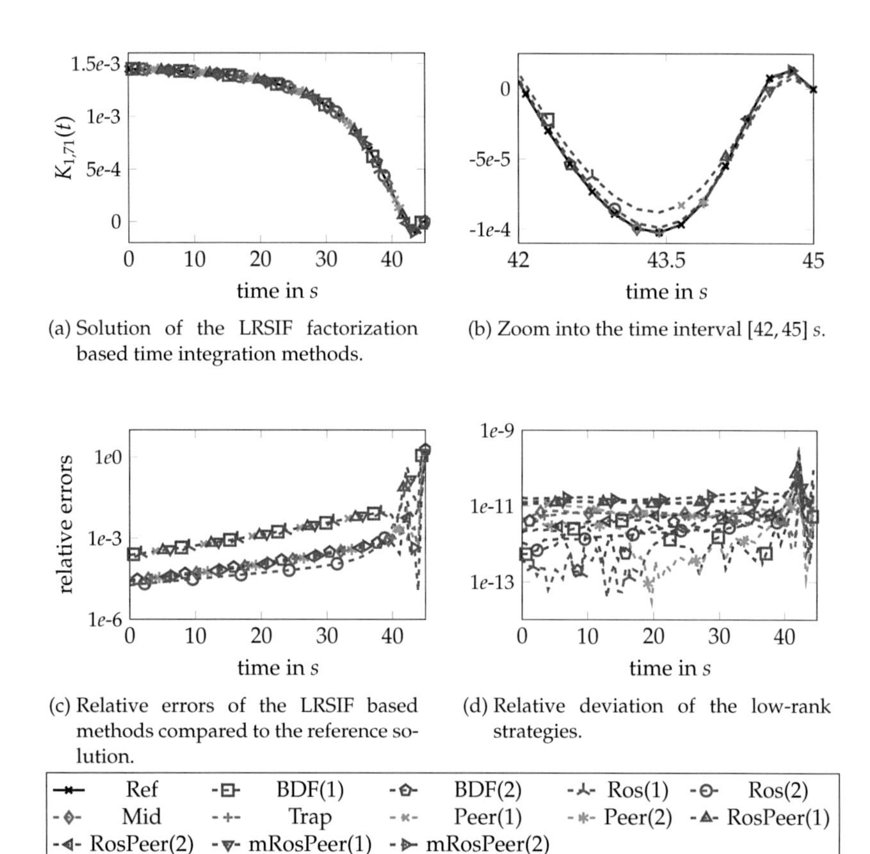

(a) Solution of the LRSIF factorization based time integration methods.

(b) Zoom into the time interval $[42, 45]$ s.

(c) Relative errors of the LRSIF based methods compared to the reference solution.

(d) Relative deviation of the low-rank strategies.

Figure 6.1.: LTI rail example: Comparison of the DME time integration methods for $n = 371$ and the time interval $[0, 45]$ s with the time step size $\tau = 0.1125$ s.

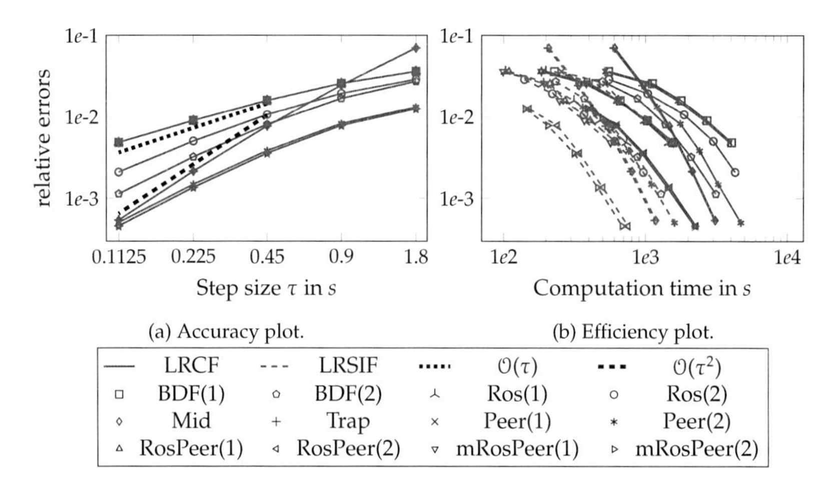

Figure 6.2.: LTI rail example: Relative error vs step size (a) and computational time (b).

Method	Solution time in s LRCF	LRSIF	Speedup	Rel. Frobenius err. LRCFvsLRSIF	LRSIFvsRef
BDF(1)	4 042.11	1 559.36	2.59	1.42e-12	4.84e-03
BDF(2)	3 124.50	1 287.00	2.43	4.83e-13	1.14e-03
Ros(1)	1 458.66	734.66	1.99	4.26e-13	4.84e-03
Ros(2)	4 286.59	965.65	4.44	3.19e-13	2.10e-03
Mid	3 098.55	1 181.12	2.62	1.11e-12	5.43e-04
Trap	3 058.95	1 152.27	2.65	9.74e-13	5.43e-04
Peer(1)	4 217.70	1 591.24	2.65	2.11e-12	4.84e-03
Peer(2)	5 029.50	1 720.43	2.92	5.63e-13	5.08e-04
RosPeer(1)	1 552.96	633.03	2.45	6.85e-12	4.84e-03
RosPeer(2)	2 274.63	749.78	3.03	8.81e-13	4.61e-04
mRosPeer(1)	1 578.08	635.48	2.48	1.23e-11	4.84e-03
mRosPeer(2)	2 245.33	743.03	3.02	3.07e-12	4.61e-04

Table 6.2.: LTI rail example: Computational times of the the low-rank factorizations, the acceleration rate, relative deviation of the low-rank strategies, and relative errors for the LRSIF methods compared to the reference.

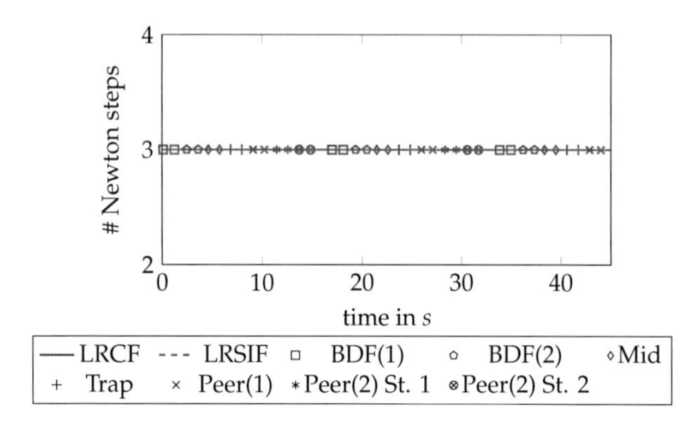

Figure 6.3.: LTI rail example: Evolution of the required number of Newton steps.

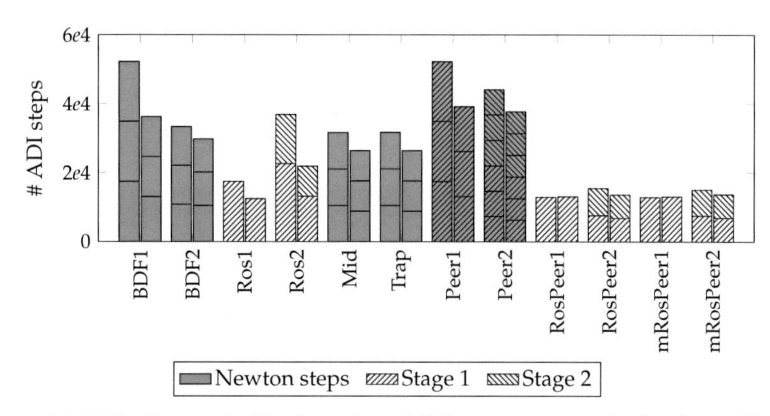

Figure 6.4.: LTI rail example: Total number of ADI steps over time for the classical low-rank DRE solvers (left bars) and low-rank symmetric solution methods (right bars). The horizontal separations represent the several Newton steps (BDF, Mid, Trap), the different stages (Ros, RosPeer, mRosPeer) and a mixture of both (Peer(2)).

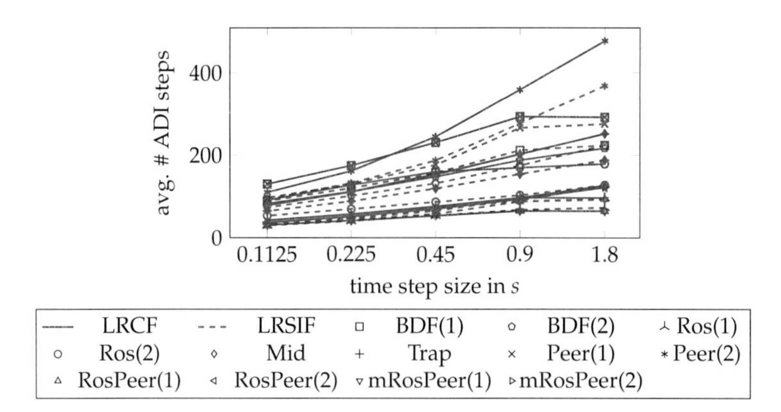

Figure 6.5.: LTI rail example: Average number of the required ADI steps per time step for different step sizes.

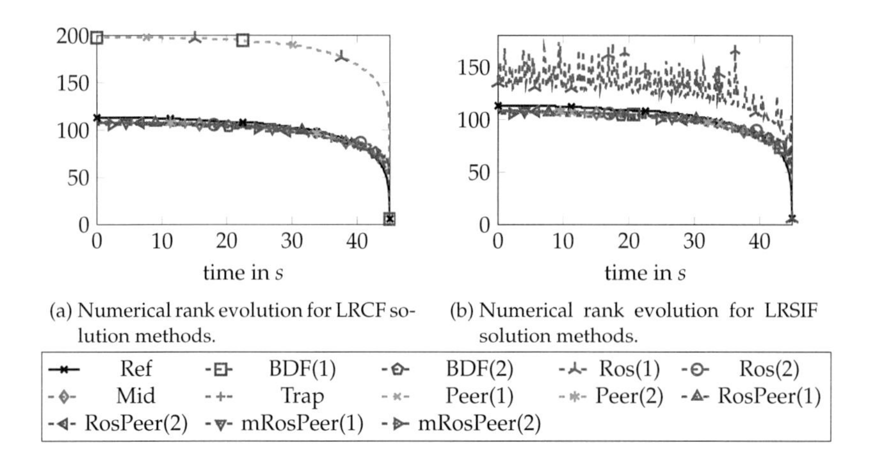

(a) Numerical rank evolution for LRCF solution methods.

(b) Numerical rank evolution for LRSIF solution methods.

Figure 6.6.: LTI rail example: Comparison of the numerical ranks.

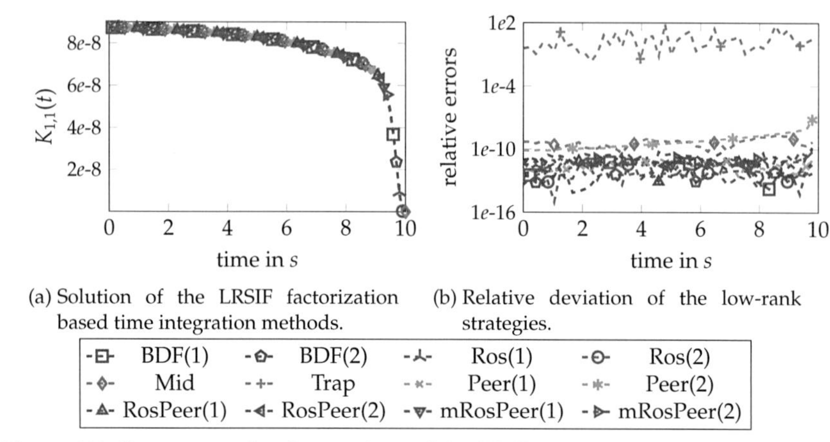

(a) Solution of the LRSIF factorization based time integration methods.

(b) Relative deviation of the low-rank strategies.

Figure 6.7.: Carex example: Comparison of the DME time integration methods for $n = 1000$ and the time interval $[0, 10]$ s with the time step size $\tau = 1/24$ s.

Carex Example

The second example originates from the CAREX benchmark collection for continuous-time algebraic Riccati equations [1, Example 4.2]. The model is a single-input-single-output (SISO) state-space system with $E, A \in \mathbb{R}^{n \times n}$, $B = C^T \in \mathbb{R}^n$. Here, we consider a system dimension of $n = 1\,000$ and the time interval $[0, 10]$ s with a constant time step size of $\tau = 1/24$ s, yielding 240 time steps to perform.

For this example no reference solution was computed. Thus, Figure 6.7 shows the evolution of the feedback component $K_{1,1}$ computed by the symmetric indefinite factorization based GDRE solvers (a) and the relative deviation of both low-rank strategies (b). As for the previous example, the computational time, the acceleration rate between both low-rank formulations and the relative deviation in the Frobenius norm are given in Table 6.3. For this example, the best performance regarding the computational time is achieved by the LRCF based RosPeer(1) method. Further, the LRCF second-order integration schemes, dealing with complex data and arithmetic, result in significantly higher computation times. For the first-order methods some of the LRCF schemes show better performance compared to the LRSIF schemes. This is due to the fact, that for first-order integrators, the right hand sides are positive definite and therefore no complex data appears in the factorizations. Further, recall from the different column compression techniques in Section 6.3, that in the case of real right hand sides, the LRCF version is computationally less expensive. Then, for the SISO system at hand, the savings regarding the columns in the right hand side factors in the LRSIF approaches cannot compensate the extra cost for the compression.

From the LRSIF solution trajectories (Figure 6.7(a)) and the relative deviation (Fig-

Method	Solution time in s		Speedup	Rel. Frobenius err.
	LRCF	LRSIF		LRCFvsLRSIF
BDF(1)	348.87	351.17	0.99	1.92e-12
BDF(2)	2 287.86	355.80	6.43	2.81e-13
Ros(1)	248.55	242.60	1.02	1.98e-12
Ros(2)	3 298.07	312.42	10.56	1.36e-14
Mid	2 308.16	224.84	10.27	5.01e-12
Trap	2 404.08	211.63	11.36	9.95e-01
Peer(1)	349.23	282.45	1.24	2.06e-12
Peer(2)	3 810.02	685.16	5.56	1.36e-09
RosPeer(1)	86.89	260.06	0.33	2.02e-13
RosPeer(2)	3 402.53	557.82	6.10	1.93e-12
mRosPeer(1)	101.45	257.62	0.39	2.11e-13
mRosPeer(2)	3 383.31	471.33	7.18	5.47e-13

Table 6.3.: Carex example: Computational times of the the low-rank factorizations, the acceleration rate, and the relative deviation of the low-rank strategies.

ure 6.7(b)) an erroneous behavior of the LRCF Trapezoidal rule can be observed. This may be caused by convergence problems within the ADI method, indicated by the large number of ADI steps taken at every Newton step. The total number of ADI steps performed over the entire simulation horizon, within the several Newton steps and/or stages of the integration schemes are presented in Figure 6.9. Excluding the LRCF Trap scheme, the required Newton steps, presented in Figure 6.8, are, again, observed to be invariant with respect to the integration method and the applied factorization. The propagation of the average number of ADI steps needed per time step is given in Figure 6.10. Again, similar to the first example, a decreasing number of ADI steps can be observed for decreasing time step sizes τ. Still, for this example, the degression is only barely mentionable for most of the integration schemes. It can also be observed that the Midpoint and Trapezoidal rules show an increasing number of average ADI steps for the step size $\tau = \frac{1}{24}$ s. By further investigations, these phenomena could be traced back to convergence problems and thus yielded unreliable results.

In Figure 6.11 the numerical ranks obtained by the LRCF (Figure 6.11(a)) and the LRSIF methods (Figure 6.11(b)) are compared. For the classical low-rank schemes we observe that every first-order method results in larger numerical ranks compared to the second-order schemes, whereas the results for the LRCF RosPeer(1) and mRosPeer(1) are slightly smaller than those of the other first-order schemes. Moreover, the Mid and Trapezoidal integrators show high oscillations and also distinguish from the other second-order schemes. In case of the symmetric indefinite ODE solvers, again, slight oscillations for the Mid and Trap schemes are visible and as for the first example, the Ros(1) LRSIF integrator shows a highly oscillatory behavior. The other LRSIF based methods achieve similar results that at the same time are comparable to the non-oscillating second-order LRCF results.

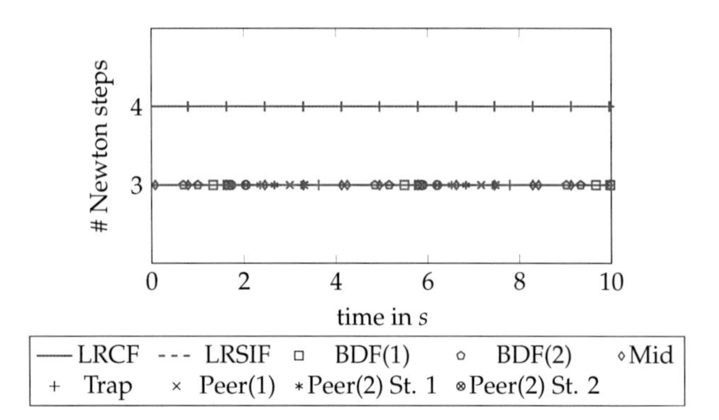

Figure 6.8.: Carex example: Evolution of the required number of Newton steps.

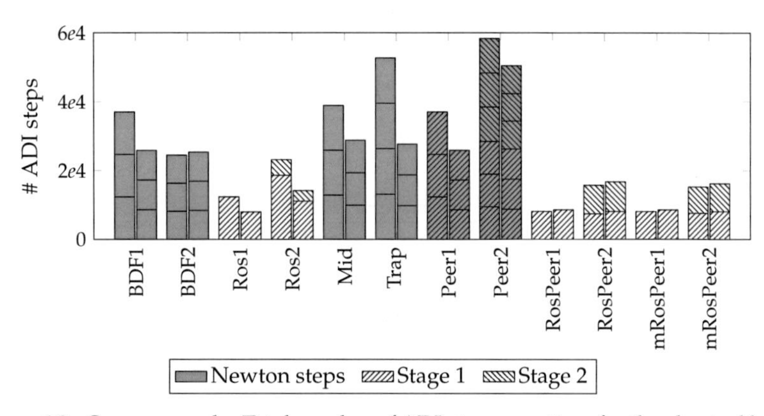

Figure 6.9.: Carex example: Total number of ADI steps over time for the classical low-rank DRE solvers (left bars) and low-rank symmetric solution methods (right bars). The horizontal separations represent the several Newton steps (BDF, Mid, Trap), the different stages (Ros, RosPeer, mRosPeer) and a mixture of both (Peer(2)).

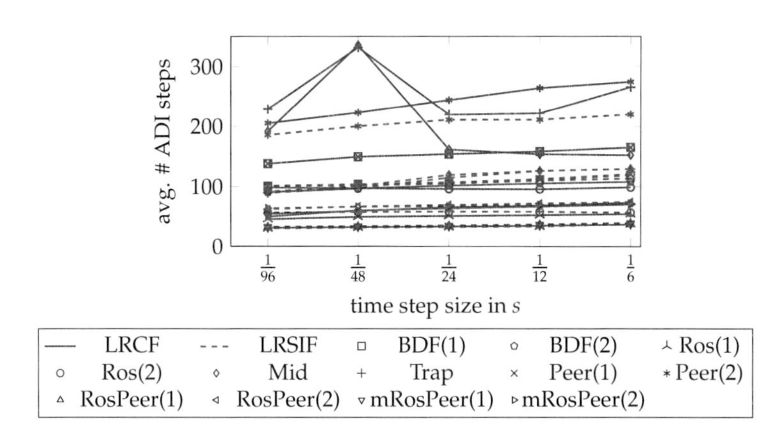

Figure 6.10.: Carex example: Average number of the required ADI steps per time step for different step sizes.

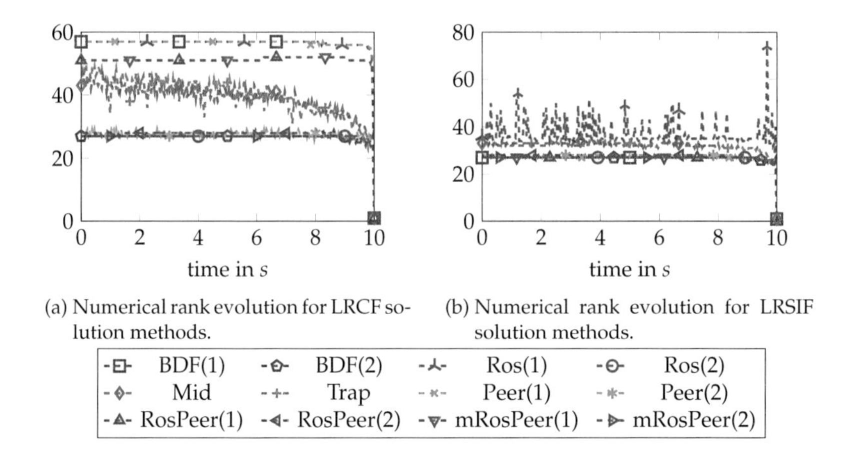

(a) Numerical rank evolution for LRCF solution methods.

(b) Numerical rank evolution for LRSIF solution methods.

Figure 6.11.: Carex example: Comparison of the numerical ranks.

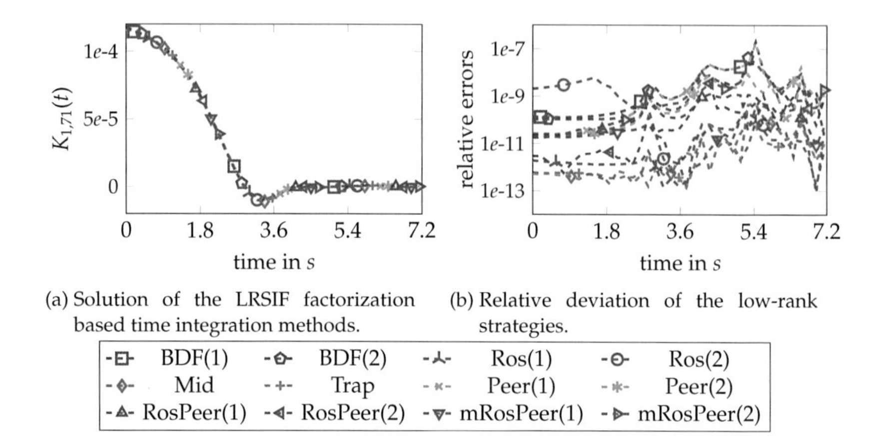

(a) Solution of the LRSIF factorization based time integration methods.

(b) Relative deviation of the low-rank strategies.

Figure 6.12.: LTV rail example: Comparison of the DME time integration methods for $n = 1357$ and the time interval $[0, 7.2]$ s with the time step size $\tau = 0.0375\,s$.

6.4.2. Non-Autonomous Control Systems

LTV Steel Rail Profile

Here, again we consider the steel rail profile example with the artificial time variability introduced in Section 4.3.3. The simulation interval $[0, 7.2]$ s is investigated with an equidistant time discretization of 192 steps based on the time step size $\tau = 0.0375\,s$. The evolution trajectories of the feedback component $K_{1,71}$ are given in Figure 6.12(a) and the relative deviation of the LRCF and LRSIF strategies are presented in Figure 6.12(b). The timings and the relative deviation in the Frobenius norm are listed in Table 6.4.

Regarding the computation times, we observe that the LRCF Rosenbrock method of second order, the Midpoint and Trapezoidal rule yield better results compared to their LRSIF counterparts that can be traced to the number of ADI steps taken, see Figure 6.14. For the other integration methods the qualitatively expected acceleration of the LRSIF based schemes is confirmed. For this example, according to the total numbers of ADI steps taken, the RosPeer(1) ans mRosPeer(1) methods followed by the RosPeer(2) schemes yield the best performance with respect to the computational times. Figure 6.13 shows the evolution of the required Newton steps that, again, is invariant with respect to the chosen methods. Figure 6.15 presents the average number of ADI steps taken per time step. As for the same arguments as in the previous examples, the number of ADI steps decreases for decreasing time step sizes τ.

Finally, the numerical ranks of the LRCF based solution factors are presented in

Figure 6.16(a). Here, the results can be clustered in three categories. The largest rank trajectories, ending at a numerical rank greater then 200 at t_0 (solve GDRE backwards in time) are obtained by the BDF(1), Ros(1) and Peer(1) methods. Then, all non-peer second-order schemes, namely the BDF(2), Ros(2), Mid and Trap schemes end up at a rank of around 130. The third category contains all peer schemes except for the Peer(1) integrator. These ODE solvers yield solution ranks of around 110 at $t = t_0$. For the LRSIF schemes, again, large deviations can be observed for the Mid and Trap solvers, see Figure 6.16(b). The same holds for the LRSIF Ros(2) method. For a better visibility, in Figure 6.16(c) the remaining LRSIF based numerical ranks are depicted without these outliers.

Method	Solution time in s LRCF	LRSIF	Speedup	Rel. Frobenius err. LRCFvsLRSIF
BDF(1)	12 592.63	3 209.66	3.92	5.24e-12
BDF(2)	10 672.05	2 719.67	3.92	4.72e-12
Ros(1)	4 757.07	2 060.90	2.31	1.17e-13
Ros(2)	24 003.93	29 095.25	0.83	8.91e-11
Mid	10 512.60	15 760.05	0.67	5.96e-14
Trap	10 278.37	17 162.27	0.60	8.89e-14
Peer(1)	7 805.46	2 942.48	2.65	4.71e-12
Peer(2)	11 336.16	3 574.97	3.17	4.58e-13
RosPeer(1)	5 355.04	1 341.96	3.99	9.11e-12
RosPeer(2)	7 495.30	1 669.76	4.49	4.15e-12
mRosPeer(1)	5 587.43	1 398.94	3.99	8.47e-12
mRosPeer(2)	8 433.73	1 808.63	4.66	2.72e-12

Table 6.4.: LTV rail example: Computational times of the the low-rank factorizations, the acceleration rate, and the relative deviation of the low-rank strategies.

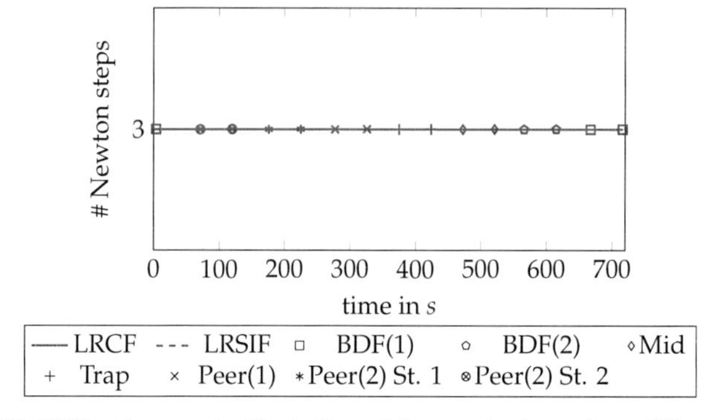

Figure 6.13.: LTV rail example: Evolution of the required number of Newton steps.

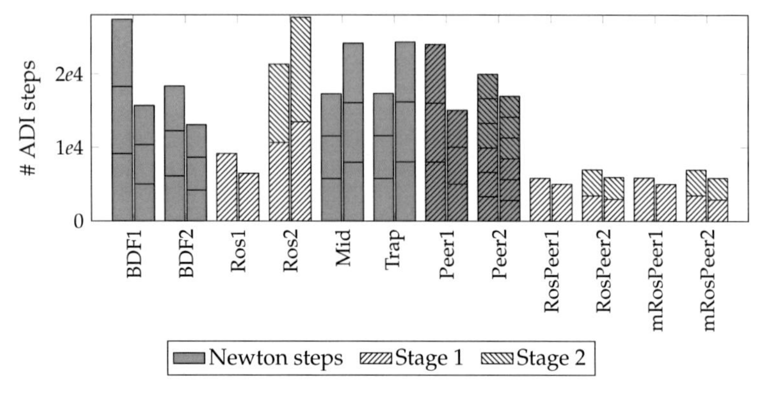

Figure 6.14.: LTV rail example: Total number of ADI steps over time for the classical low-rank DRE solvers (left bars) and low-rank symmetric solution methods (right bars). The horizontal separations represent the several Newton steps (BDF, Mid, Trap), the different stages (Ros, RosPeer, mRosPeer) and a mixture of both (Peer(2)).

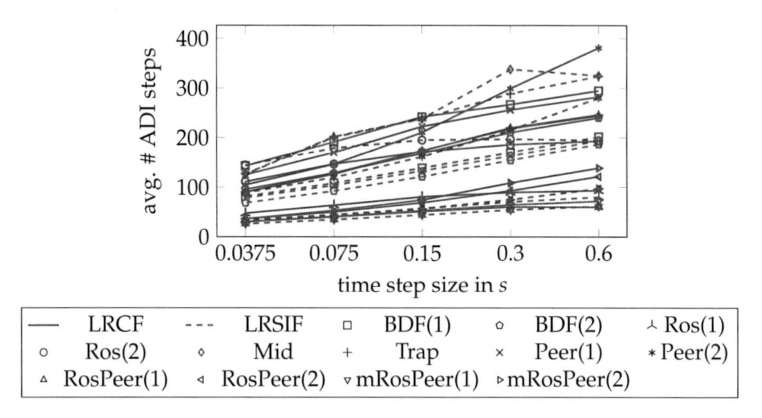

Figure 6.15.: LTV rail example: Average number of the required ADI steps per time step for different step sizes.

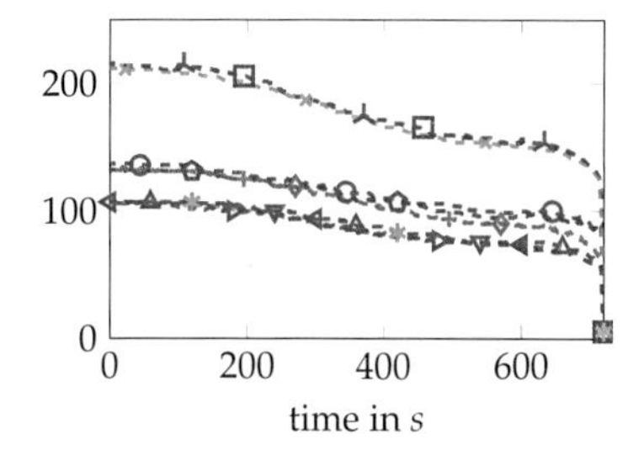

(a) Numerical rank evolution for LRCF solution methods.

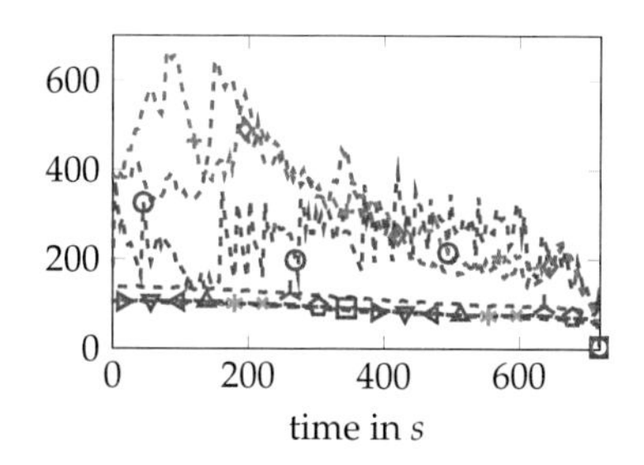

(b) Numerical rank evolution for LRSIF solution methods.

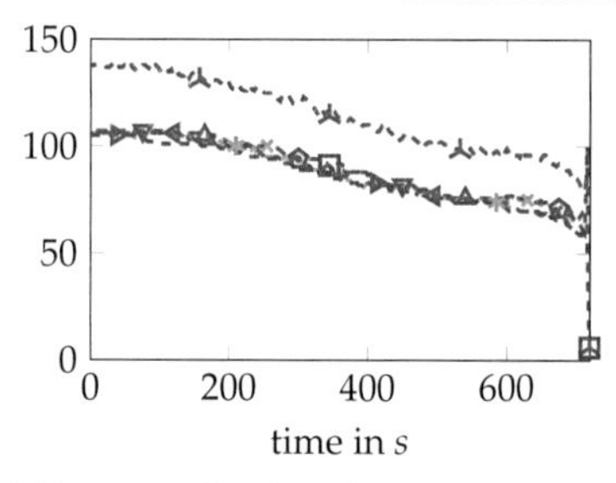

(c) Numerical rank evolution for LRSIF solution methods - without Ros(2), Mid and Trap.

-□-	BDF(1)	-⬠-	BDF(2)	-⅄-	Ros(1)	-⊖-	Ros(2)
-◇-	Mid	-+-	Trap	-✳-	Peer(1)	-✳-	Peer(2)
-▲-	RosPeer(1)	-◀-	RosPeer(2)	-▼-	mRosPeer(1)	-▶-	mRosPeer(2)

Figure 6.16.: LTV rail example: Comparison of the numerical ranks.

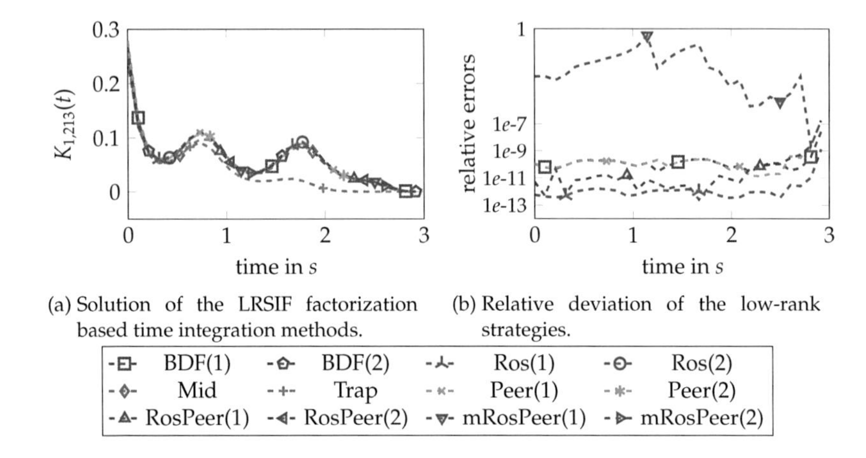

(a) Solution of the LRSIF factorization based time integration methods.

(b) Relative deviation of the low-rank strategies.

Figure 6.17.: Burgers example: Comparison of the DME time integration methods for $n = 1500$ and the time interval $[0, 3]$ s with the time step size $\tau = 1/48$ s.

Burgers Equation Example

The final model is, again, the Burgers equation example, presented in Section 4.3.2. Here, a spatial dimension of $n = 1\,500$ and a time interval of $[0, 3]$ s with a time step size of $\tau = \frac{1}{48}$ s, resulting in 144 steps, are considered. For this example, complex ADI shifts occur within the Lyapunov solver. That is, as described in Section 6.1.6, the second-order time integration schemes based on the LRCF approach cannot be used straightforwardly. Thus, only first-order results are presented. In the same manner as before, the solution trajectories and the associated relative deviation for the remaining methods are given in Figures 6.17(a) and 6.17(b). The computational times and the relative deviation of the two factorization strategies are given in Table 6.5.

Again, the RosPeer(1) method achieves the best performance with respect to the timings, whereas the LRCF version is even faster compared to the LRSIF counterpart. This is indicated by the total number of ADI steps taken, presented in Figure 6.19. In contrast to the previous examples, the number of Newton steps slightly varies, see Figure 6.18. In particular the LRSIF versions of the Midpoint and Trapezoidal rules, as well as the two stages of the Peer(2) scheme reveal some convergence difficulties that at least can be seen in the timings of the Mid and Trap integrators. Except for these outliers, the number of required Newton steps is still observed to be invariant with respect to the integration schemes and the applied low-rank strategy. Finally, in Figure 6.20 the average number of required ADI steps is presented. As for the Carex example that is also a SISO model, the decrease of average ADI steps for decreasing step sizes, is also of minor significance.

For this example, the numerical ranks are depicted in Figure 6.21. Here, the LRCF based methods (all first-order) all show a similar rank evolution, see Figure 6.21(a). For the LRSIF schemes, again the Mid and Trap methods strongly differ from the other methods. In particular, a monotonic growth can be observed that explains the high computation times for these methods. As in the LTI Rail and Carex examples, the numerical rank trajectory obtained by the LRSIF Ros(1) method is significantly larger than those of the other methods (excluding Mid and Trap) and in addition highly oscillates, see Figure 6.21(d). The LRSIF based numerical ranks, excluding the Ros(1), Ros(2), Mid, and Trap ODE solvers, are given in 6.21(c).

Method	Solution time in s LRCF	LRSIF	Speedup	Rel. Frobenius err. LRCFvsLRSIF
BDF(1)	1 428.09	258.27	5.53	1.65e-10
BDF(2)	-	260.45	-	-
Ros(1)	901.43	544.21	1.66	9.71e-13
Ros(2)	-	2 622.64	-	-
Mid	-	4 960.12	-	-
Trap	-	7 010.54	-	-
Peer(1)	934.94	200.81	4.66	1.79e-10
Peer(2)	-	448.64	-	-
RosPeer(1)	82.23	167.06	0.49	5.07e-11
RosPeer(2)	-	401.24	-	-
mRosPeer(1)	390.13	170.03	2.29	3.22e-02
mRosPeer(2)	-	485.42	-	-

Table 6.5.: Burgers example: Computational times of the the low-rank factorizations, the acceleration rate, and the relative deviation of the low-rank strategies.

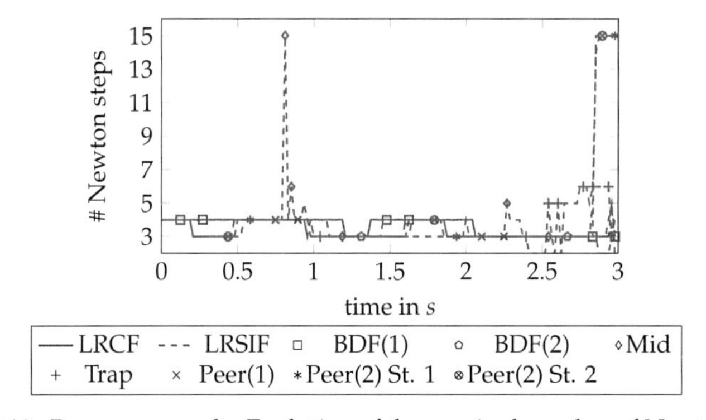

Figure 6.18.: Burgers example: Evolution of the required number of Newton steps.

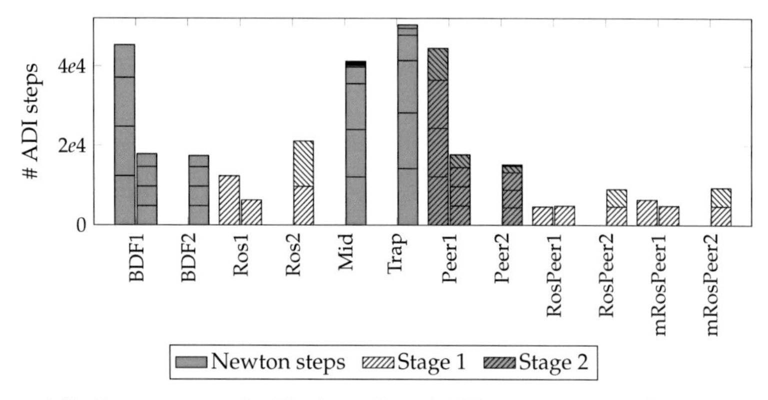

Figure 6.19.: Burgers example: Total number of ADI steps over time for the classical low-rank DRE solvers (left bars) and low-rank symmetric solution methods (right bars). The horizontal separations represent the several Newton steps (BDF, Mid, Trap), the different stages (Ros, RosPeer, mRosPeer) and a mixture of both (Peer(2)).

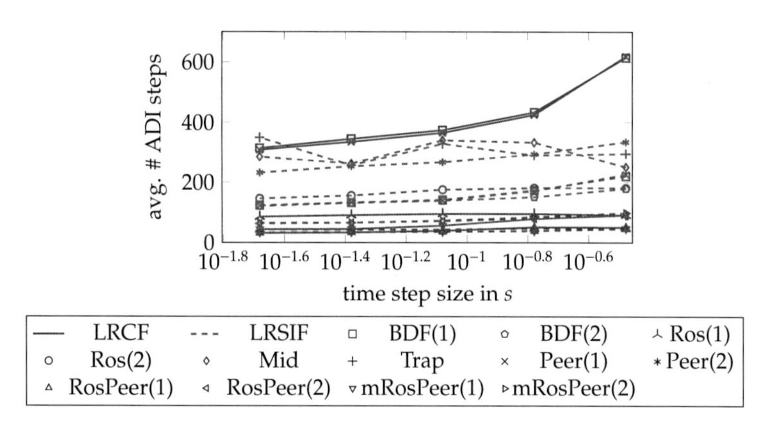

Figure 6.20.: Burgers example: Average number of the required ADI steps per time step for different step sizes.

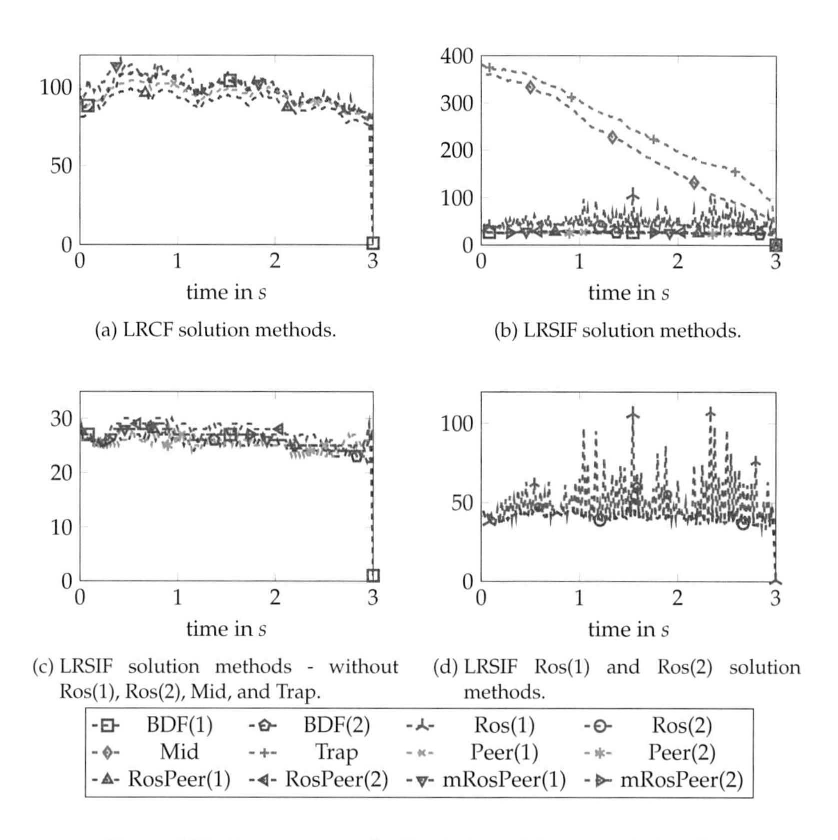

Figure 6.21.: Burgers example: Evolution of the numerical ranks.

6.5. Summary and Conclusions

In this chapter, the numerical solution of differential matrix equations based on low-rank solution strategies is presented. The differential Lyapunov equation has been considered in Section 4.3.2 already, and since it is a special case of the differential Riccati equation, here the investigations are restricted to the Riccati case. The classical two-term and the symmetric indefinite factorizations, applied to the BDF, Rosenbrock methods, the Midpoint and Trapezoidal rules, and peer methods are compared. For an integration order of $s \geq 2$, the right hand side of the algebraic Lyapunov equations, arising in the innermost iteration of the presented integration methods, become indefinite and therefore the LRCF solution methods have to deal with complex data and arithmetic. Moreover, it is shown that for complex shifts, entering the solution methods for the Lyapunov equations, the LRCF based integration schemes cannot be applied straightforwardly. In case of the LRSIF based schemes it can be avoided that the factors become complex and at the same time the column size of these factors can in advance analytically be reduced. In Tables 6.6 and 6.7, the achieved column sizes of the right hand side factors G for the non-autonomous and autonomous (G)DRE, for both the classical and symmetric indefinite factorization, are summarized. Tables 6.8 and 6.9 summarize the corresponding results for (G)DLE. Note that the presented tables only give the absolute column sizes and do not reveal if the factors are real- or complex-valued. Moreover, in the numerical experiments it turned out to be needful to apply the compression procedure to every right hand side factor G_k and every computed solution factor Z_k, L_k, respectively at every time step of the integration methods. Thus, the column size of G_k is a major issue regarding the computational effort of the presented solution methods for the solution of the (G)DREs and (G)DLEs.

The corresponding column compression techniques are presented for the classical low-rank factorization for real and complex data, as well as for the symmetric indefinite case. Since the rank-revealing compression methods are applied to the solution factors at every time step, the column sizes of the factors at the same time represent the numerical rank of the computed solutions. It should also be mentioned that the complex column compression for the LRCF methods and the compression technique for the LRSIF approach are of the same computational complexity, exceeding that of the real LRCF compression. For that reason, using first-order time integration methods, resulting in definite right hand sides, applied to a SISO system, the a-priori column size reduction within the LRSIF approaches cannot compensate the higher cost of the compression compared to the LRCF version.

From the numerical experiments, it has been observed, that in most cases, the LRCF BDF(1), Ros(1) and Peer(1) methods result in considerably larger ranks than the other methods. Further, in case of the LRSIF schemes the Ros(1) scheme struggled with high oscillations regarding the rank trajectories. Anyway, the most suspicious behavior was encountered for the Mid and Trap schemes. For the Carex example an erroneous behavior for the LRCF formulation and oscillating rank trajectories for both low-rank strategies were observed. For the Rail LTV example the ranks for the LRSIF Mid and Trap schemes showed high deviations that consequently resulted in significantly high computation times. For the Burgers example only the LRSIF formulation of the Mid and Trap methods could be investigated that again yielded high numerical ranks in close coherence with very high computational times compared to the other methods. Moreover, note that the results for the numerical ranks depend on the column compression technique and the associated truncation tolerance used throughout the computations.

Summarizing, it should be mentioned that for some of the examples, single methods of both strategies struggled with convergence problems with respect to either Newton's method or the ADI iteration, resulting in an unpredictable behavior of the solutions in terms of, e.g., the rank evolution. Further, this may lead to bad computational times and even unreliable or wrong solutions.

In general, the numerical experiments have proven the theoretically predicted better performance of the LRSIF based methods. As a very positive outcome, it is noteworthy that the RosPeer and modified RosPeer methods in most applications show best performance with respect to the computational times and achieved accuracy.

	Column size of G_k	
	Classical	Symm. Indef.
s-step BDF	$\sum_{j=1}^{s} n_{Z_{k+1-j}} + q + m$	$\sum_{j=1}^{s} n_{L_{k+1-j}} + q + m$
Ros(1)	$n_{Z_k} + q + m$	$n_{L_k} + q$
Ros(2) 1st	$3n_{Z_k} + q + m$	$q + 2n_{L_k}$
Ros(2) 2nd	$3n_{Z_k} + 4n_{T_1} + q + 5m$	$2n_{L_k} + 2n_{T_1} + q$
Midpoint	$3n_{Z_k} + q + 2m$	$2n_{L_k} + q + m$
Trapezoidal	$3n_{Z_k} + 2(q + m)$	$2n_{L_k} + 2q + m$
Peer	$4\sum_{j=1}^{s} n_{Z_{k-1,j}} + 3\sum_{j=1}^{i-1} n_{Z_{k,j}} + (s + i)(q + m)$	$3\sum_{j=1}^{s} n_{L_{k-1,j}} + 2\sum_{j=1}^{i-1} n_{L_{k,j}} + (s + i)q + m$
RosPeer	$3\sum_{j=1}^{s-1} n_{Z_{k-1,j}} + n_{Z_{k-1,s}} + 3\sum_{j=1}^{i-1} n_{Z_{k,j}} + sq + (s+2)m$	$2\sum_{j=1}^{s-1} n_{L_{k-1,j}} + n_{L_k} + 2\sum_{j=1}^{i-1} n_{L_{k,j}} + sq + 2m$
mRosPeer	$3\sum_{j=1}^{s}\left(n_{Z_{k-1,j}} + n_{\hat{Z}_{k-1,j}}\right) + \sum_{j=1}^{i-1} n_{\hat{Z}_{k,j}} + sq + (s+2)m$	$2\sum_{j=1}^{s}(n_{\hat{L}_{k-1,j}} + n_{L_{k-1,j}}) + \sum_{j=1}^{i-1} n_{L_{k,j}} + sq + 2m$

Table 6.6.: Non-autonomous DRE: Column size of the right hand side factors G for the classical and the symmetric indefinite factorization.

	Column size of G_k	
	Classical	Symm. Indef.
Ros(2) 1st	$3n_{Z_k} + q + m$	$2n_{L_k} + q$
Ros(2) 2nd	$n_{T_1} + m$	n_{T_1}
Trapezoidal	$3n_{Z_k} + q + 2m$	$2n_{L_k} + q + m$
Peer	$4\sum_{j=1}^{s} n_{Z_{k-1,j}} + 3\sum_{j=1}^{i-1} n_{Z_{k,j}} + q + (s + i)m$	$3\sum_{j=1}^{s} n_{L_{k-1,j}} + 2\sum_{j=1}^{i-1} n_{L_{k,j}} + q + m$
RosPeer	$\sum_{j=1}^{s} n_{Z_{k-1,j}} + 3\sum_{j=1}^{i-1} n_{Z_{k,j}} + q + (s + 2)m$	$\sum_{j=1}^{s} n_{L_{k-1,j}} + 2\sum_{j=1}^{i-1} n_{L_{k,j}} + q + 2m$
mRosPeer	$\sum_{j=1}^{s} n_{\hat{Z}_{k-1,j}} + \sum_{j=1}^{i-1} n_{\hat{Z}_{k,j}} + q + (s + 2)m$	$\sum_{j=1}^{s} n_{\hat{L}_{k-1,j}} + \sum_{j=1}^{i-1} n_{\hat{L}_{k,j}} + q + (s + 1)m$

Table 6.7.: Autonomous DRE: Column size of the right hand side factors G for the classical and the symmetric indefinite factorization.

	Column size of G_k	
	Classical	Symm. Indef.
s-step BDF	$\sum_{j=1}^{s} n_{Z_{k+1-j}} + m$	$\sum_{j=1}^{s} n_{L_{k+1-j}} + m$
Ros(1)	$n_{Z_k} + m$	$n_{L_k} + m$
Ros(2) 1st	$3n_{Z_k} + m$	$2n_{L_k} + m$
Ros(2) 2nd	$3n_{Z_k} + 4n_{T_1} + m$	$2n_{L_k} + 2n_{T_1} + m$
Midpoint	$3n_{Z_k} + m$	$2n_{L_k} + m$
Trapezoidal	$3n_{Z_k} + 2m$	$2n_{L_k} + 2m$
Peer	$4\sum_{j=1}^{s} n_{Z_{k-1,j}} + 3\sum_{j=1}^{i-1} n_{Z_{k,j}} + (s+i)m$	$3\sum_{j=1}^{s} n_{L_{k-1,j}} + 2\sum_{j=1}^{i-1} n_{L_{k,j}} + (s+i)m$
RosPeer	$3\sum_{j=1}^{s-1} n_{Z_{k-1,j}} + n_{Z_{k-1,s}} + 3\sum_{j=1}^{i-1} n_{Z_{k,j}} + sm$	$2\sum_{j=1}^{s-1} n_{L_{k-1,j}} + n_{L_{k-1,s}} + 2\sum_{j=1}^{i-1} n_{L_{k,j}} + sm$
mRosPeer	$3\sum_{j=1}^{s}\left(n_{\hat{Z}_{k-1,j}} + n_{Z_{k-1,j}}\right) + \sum_{j=1}^{i-1} n_{\hat{Z}_{k,j}} + sm$	$2\sum_{j=1}^{s}\left(n_{\hat{L}_{k-1,j}} + n_{L_{k-1,j}}\right) + \sum_{j=1}^{i-1} n_{\hat{L}_{k,j}} + sm$

Table 6.8.: Non-autonomous DLE: Column size of the right hand side factors G for the classical and the symmetric indefinite factorization.

	Column size of G_k	
	Classical	Symm. Indef.
Ros(2) 1st	$3n_{Z_k} + m$	$2n_{L_k} + m$
Ros(2) 2nd	n_{T_1}	n_{T_1}
Trapezoidal	$3n_{Z_k} + m$	$2n_{L_k} + m$
Peer	$4\sum_{j=1}^{s} n_{Z_{k-1,j}} + 3\sum_{j=1}^{i-1} n_{Z_{k,j}} + m$	$3\sum_{j=1}^{s} n_{L_{k-1,j}} + 2\sum_{j=1}^{i-1} n_{L_{k,j}} + m$
RosPeer	$\sum_{j=1}^{s} n_{Z_{k-1,j}} + 3\sum_{j=1}^{i-1} n_{Z_{k,j}} + m$	$\sum_{j=1}^{s} n_{L_{k-1,j}} + 2\sum_{j=1}^{i-1} n_{L_{k,j}} + m$
mRosPeer	$\sum_{j=1}^{s} n_{\hat{Z}_{k-1,j}} + \sum_{j=1}^{i-1} n_{\hat{Z}_{k,j}} + m$	$\sum_{j=1}^{s} n_{\hat{L}_{k-1,j}} + \sum_{j=1}^{i-1} n_{\hat{L}_{k,j}} + m$

Table 6.9.: Autonomous DLE: Column size of the right hand side factors G for the classical and the symmetric indefinite factorization.

CHAPTER 7 CONCLUSIONS AND OUTLOOK

Contents

7.1. Summary and Conclusions

In this thesis numerical methods for large-scale linear control systems, with focus on the time-varying case, have been studied. The focal topics are the finite-time optimal control [143] and model order reduction of linear time-varying dynamical systems. Different approaches using SLS [141] or parametric model order reduction [21] based on LTI model approximations and a direct application of balanced truncation for LTV systems [183, 202, 171] are considered. In both major fields of application, as a main step, differential matrix equations have to be solved. The solution of these differential equations can be reduced to the solution of a number of algebraic Lyapunov equations. Motivated by the practically observed and theoretically predicted fast singular value decay of the solutions [159, 9, 198, 19, 20], efficient implementations of the low-rank factorization based ADI [23, 24, 25, 126] and Krylov subspace methods [185, 61] are reviewed. In this thesis, two different low-rank factorization strategies are distinguished. Due to cyclically growing factors and the appearance of complex data for some of the low-rank schemes, specially adjusted rank-revealing column compression techniques are presented. Extensive numerical tests show the performance of the presented numerical methods.

In Chapter 3 a finite-time tracking type optimal control problem with application to an inverse heat conduction problem is considered. In order to later on classify the optimal control formulation, a short overview on inverse problems is presented. Then, as a first step the underlying heat conduction problem is formulated in terms of a linear time-invariant control system. The procedure aims at finding an optimal control that drives the system outputs to a given set of measurements. Based on the given PDE together with the boundary conditions, the resulting generalized state-space system turned out to be inhomogeneous. The well-known Hamilton-Jacobi theory, reviewed in Section 2.1.5, is applied to the inhomogeneous generalized system. As the outcome, for the solution of the given tracking-type optimal control problem and therefore of the inverse problem, a differential Riccati equation and the adjoint state-space system have to be solved. Numerical tests, showing the efficiency of the methods, are presented. First, an artificial test example is investigated with respect to the behavior of the reconstructed input signal based on the number of measurements given, the choice of the weighting matrices within the optimal control performance index, and the quality of the underlying heat model in terms of a perturbation applied to the heat transfer coefficient within the boundary conditions. The second example describes a simplified real-world example based on measurements recorded within a real heating process. Given the reference input in case of the first example, a relative error of the reconstructed input signal within a single-digit percent range or even the range of one per thousand could be obtained. Due to the absence of a reference solution for the real-world application, the relative error of the computed outputs compared to the measurements was presented that also lay in the range of one per thousand.

Three different approaches to MOR for time-varying control systems are presented in Chapter 4. An SLS approach using LTI subsystems was proposed for the approximation of the time-variability. Then, each of the subsystems was reduced by the balanced truncation method. For the second procedure, the time-dependency is interpreted as a parameter and the PMOR approach given in [14] is applied to the resulting LTI systems, defined for a set of a-priori chosen parameter sample points. The last MOR scheme directly applies the BT method defined for LTV systems. Here, the main ingredient is the solution of the differential reachability and observability Lyapunov equations. Due to the numerical complexity of the latter MOR ansatz, the experiments were divided into two parts. The first of which compares the SLS and PMOR approaches applied to a moving load problem given by a machine-stand-slide-structure, where a moving tool slide induces a certain thermal input to the stand system. Considering a comparable reduced dimension, both reduction schemes result in the same ranges for the relative reduction errors, whereas the PMOR approach shows the better offline timings. On the other hand, the SLS ansatz is slightly advantageous with respect to the online simulation that is due to the LU factorizations that can be reused within the forward time integration method. A comparison of all three MOR procedures is given for the artificially constructed LTV rail example. Here, the SLS and PMOR schemes show a similar behavior as for the stand-slide model. Further, it can be observed that the BT method for LTV systems based on time-varying truncation matrices cannot compete with the other methods with respect to the offline and online timings. Still, the relative reduction errors outperform the SLS and PMOR results. Moreover, using global truncation matrices, similar to the PMOR approach, the BT scheme for LTV systems can catch up with the simulation timings for the reduced-order models at the cost of a slightly increased offline time caused by the additional computation of the global projectors. At the same time, the relative reduction errors still top the results obtained by the SLS and PMOR schemes.

A number of implicit time integration schemes necessary for the solution of the differential Riccati and Lyapunov equations from Chapters 3 and 4 are presented in Chapter 5. Here, the BDF, classical Rosenbrock, Midpoint and Trapezoidal rules, implicit peer methods, as well as the Rosenbrock-type peer methods are investigated for the application to differential matrix equations. For the latter, also a modification avoiding a number of applications of the Jacobian is proposed.

The subsequent Chapter 6 deals with the classical low-rank Cholesky-type and symmetric indefinite factorization based formulations of the previously introduced matrix-valued ODE solvers. In particular, the block sizes of the right hand side factors, directly influencing the performance of the Lyapunov solvers, are emphasized. For the LRCF schemes it is revealed that all implicit methods of integration order ≥ 2 result in complex data and therefore require complex arithmetic. Moreover, it is shown that for complex shifts, occurring in the solution methods for the Lyapunov equations, the LRCF based integrators of order ≥ 2 can no longer be applied straightforwardly. For the LRSIF methods these drawbacks and thus complex data and arithmetic are avoided. In case of complex shifts the realification strategies introduced for the LRCF ADI method can be applied. That is, in any possible configuration, the computations stay real. At the same time the crucial block sizes of the right hand sides can be analytically reduced a-priori by eliminating redundant information, not visible for the LRCF decompositions. Further, factorization dependent column compression techniques are proposed. That is, the cases of an LRCF factorization dealing with real or complex data and an LRSIF factorization are distinguished and separately handled by the respective rank-revealing compression. From the compression procedures it can be seen that the LRCF compression for real factors is of lower cost compared to the other procedures, that have the same computational complexity. For that reason, dealing with first-order integration methods, applied to SISO systems, the LRCF schemes are capable of outperforming the a-priori analytic rank reduction used in the LRSIF schemes. The numerical experiments show the performance of the LRCF and LRSIF methods. Despite the case where first-order integration methods are applied to SISO systems, the LRSIF schemes, in general, show the better performance as expected from the theoretical results. Still, there are some exceptions.

7.2. Future Research Perspectives

In general, this thesis is a mixture of problems strongly related to applications, whereas others are of more theoretical nature. The same applies to the possible future research perspectives.

In Chapter 3, we have observed that the theoretically induced drop of the optimal control u, defined by the feedback law (3.23), at the final time t_f is strongly influenced by the given measurements. Thus, the optimal placement of sensors, recording the measurement information, is an interesting question that is probably able to significantly improve the reconstruction results and therefore worth to be considered. Moreover, so far only single input LTI systems have been studied. In general, the presented framework is capable of handling LTV systems of MIMO structure. Still, the performance, in particular with respect to multiple and/or moving input signals is not yet considered. Another challenging task is the problem of reconstructing the heat transfer coefficients. In practice, these coefficients are either known approximately only or have to be determined by complex thermal models. Considering the transfer coefficient to be the searched for input signal, the resulting dynamical model becomes bilinear and thus requires to adapt the optimal control approach, see e.g., [150, 166, 107, 64].

Considering BT for LTV systems, in Chapter 4, we have encountered that the zero initial conditions of the reachability and observability DLEs yield zero truncation matrices $V(t)$, $W(t)$ for $t = t_0$ and $t = t_f$ and thus does not allow a suitable reduction of the underlying dynamical system at the boundaries of the time interval. Although we have shown that this is not a problem for $x(t_0) = 0$, the problem remains for non-zero initial conditions and $t = t_f$. In [171], the authors suggest to embed the considered interval $[t_0, t_f]$ in a larger surrogate interval $[t_0^s, t_f^s]$ with $t_0^s < t_0 < t_f < t_f^s$. Having a certain application with a specific time interval for the forward simulation in mind, this seems to be a rather random approach. Moreover, this directly demands for a suitable strategy to choose the embedding interval. Therefore, a more detailed elaboration of possible remedies seems to be worth the effort.

Concerning the high computational times for both the LRCF and LRSIF based integration methods for matrix-valued differential matrix equations, further improvements with respect to efficient solution methods are desirable. In [119] a framework using parallel computer architectures based on the Parareal method [142] for the solution of (G)DREs is presented. At the moment, this procedure cannot handle large-scale systems. That is, a low-rank formulation of the Parareal approach could drastically reduce the computation times using parallel computers and is therefore a highly desirable tool for the application to large-scale DREs, or in general DMEs. Due to the absence of running code for high-order one-step integration methods, there is also no high-order multistep or peer method available. That is, the formulation of, e.g., Rosenbrock methods of order 3 or 4 in terms of efficient low-rank strategies has to be treated in the future. One can also think of an incorporation of the splitting based ODE solvers [190, 191], providing high-order methods more easily. Further, so far, major effort has been put in the solution of the matrix-valued symmetric differential Riccati and Lyapunov equations, whereas an extension to the large-scale non-symmetric differential Sylvester equation is not yet considered.

Another critical issue is the dimension of the models. Considering models with very large dimensions easily exceeds the limits of direct solvers. Therefore, iterative solution methods need to be considered for the solution of the linear systems occurring within the presented time integration methods. In, e.g., [70, 22, 204] an inexact solution of Newton's method based on iterative methods has already been investigated for algebraic Riccati equations. An extension to differential matrix equations is a highly interesting topic to be considered in the future.

Appendix

APPENDIX A

FOURTH-ORDER ROSENBROCK METHOD

The most common Rosenbrock methods are of order 4. In [182], a fourth-order Rosenbrock scheme with 4-stages is presented. Applied to the GDRE (2.12), the procedure reads

$$\begin{aligned}
X_{k+1} =& X_k + \tau_k\left(\frac{19}{18}K_1 + \frac{1}{4}K_2 + \frac{25}{216}K_3 + \frac{125}{216}K_4\right), \\
\tilde{A}_k^T K_1 E + E^T K_1 \tilde{A}_k =& -\mathcal{R}\left(t_k, X_k\right) - \frac{1}{2}\tau_k \mathcal{R}_t(t_k, X_k), \\
\tilde{A}_k^T K_2 E + E^T K_2 \tilde{A}_k =& -\mathcal{R}\left(t_{k+1}, X_k + \tau_k K_1\right) + 4E^T K_1 E + \frac{3}{2}\tau_k \mathcal{R}_t(t_k, X_k) \\
\tilde{A}_k^T K_3 E + E^T K_3 \tilde{A}_k =& -\mathcal{R}\left(t_k + \frac{3}{5}\tau_k, X_k + \frac{24}{25}\tau_k K_1 + \frac{3}{25}\tau_k K_2\right) \\
& - \frac{185}{25}E^T K_1 E - \frac{6}{5}E^T K_2 E - \frac{121}{50}\tau_k \mathcal{R}_t(t_k, X_k) \\
\tilde{A}_k^T K_4 E + E^T K_4 \tilde{A}_k =& -\mathcal{R}\left(t_k + \frac{3}{5}\tau_k, X_k + \frac{24}{25}\tau_k K_1 + \frac{3}{25}\tau_k K_2\right) \\
& + \frac{56}{125}E^T K_1 E + \frac{27}{125}E^T K_2 E + \frac{1}{5}E^T K_3 E - \frac{29}{250}\tau_k \mathcal{R}_t(t_k, X_k)
\end{aligned} \tag{A.1}$$

with $\tilde{A}_k := \frac{1}{2}\Big(\tau_k(A_k - B_k B_k^T X_k E) - E\Big)$. Note that scheme (A.1) was developed according to the representation (5.8), set up for autonomized ODEs and thus contains the time derivative $\mathcal{R}_t(t_k, X_k)$ of the Riccati operator. Anyway, observing that the third and fourth stage of the 4-stage scheme (A.1) share the same function evaluation in their right hand sides of the algebraic Lyapunov equations, a simplified scheme of (A.1) can be formulated and reads

$$
\begin{aligned}
X_{k+1} =& X_k + \tau_k \left(\frac{19}{18} K_1 + \frac{1}{4} K_2 + \frac{25}{216} K_3 + \frac{125}{216} K_4 \right), \\
\tilde{A}_k^T K_1 E + E^T K_1 \tilde{A}_k =& - \mathcal{R}\left(t_k, X_k\right) - \frac{1}{2} \tau_k \mathcal{R}_t(t_k, X_k), \\
\tilde{A}_k^T K_2 E + E^T K_2 \tilde{A}_k =& - \mathcal{R}\left(t_{k+1}, X_k + \tau_k K_1\right) + 4 E^T K_1 E + \frac{3}{2} \tau_k \mathcal{R}_t(t_k, X_k) \\
\tilde{A}_k^T K_3 E + E^T K_3 \tilde{A}_k =& - \mathcal{R}\left(t_k + \frac{3}{5}\tau_k, X_k + \frac{24}{25}\tau_k K_1 + \frac{3}{25}\tau_k K_2\right) \\
& - \frac{185}{25} E^T K_1 E - \frac{6}{5} E^T K_2 E - \frac{121}{50} \tau_k \mathcal{R}_t(t_k, X_k), \\
\tilde{A}_k^T \tilde{K}_4 E + E^T \tilde{K}_4 \tilde{A}_k =& + \frac{981}{125} E^T K_1 E + \frac{177}{125} E^T K_2 E + \frac{1}{5} E^T K_3 E + \frac{238}{125} \tau_k \mathcal{R}_t(t_k, X_k), \\
K_4 =& \tilde{K}_4 + K_3.
\end{aligned}
\tag{A.2}
$$

In the autonomous case with constant coefficient matrices, the time variable does not explicitly affect the right hand sides of the stage equations. Therefore, the four-stage scheme can be further simplified by extracting the already computed parts of the Riccati operators from the previous stages. Moreover, we have $\mathcal{R}_t(t_k, X_k) = 0$ and one obtains the condensed scheme

$$
\begin{aligned}
X_{k+1} =& X_k + \tau_k \left(\frac{19}{18} K_1 + \frac{1}{4} K_2 + \frac{25}{216} K_3 + \frac{125}{216} K_4 \right), \\
\tilde{A}_k^T K_1 E + E^T K_1 \tilde{A}_k =& - \mathcal{R}\left(t_k, X_k\right), \\
\tilde{A}_k^T \tilde{K}_2 E + E^T \tilde{K}_2 \tilde{A}_k =& \tau_k^2 E^T K_1 B B^T K_1 E + 2 E^T K_1 E, \\
K_2 =& \tilde{K}_2 - K_1, \\
\tilde{A}_k^T \tilde{K}_3 E + E^T \tilde{K}_3 \tilde{A}_k =& - \frac{245}{25} E^T K_1 E - \frac{36}{25} E^T K_2 E \\
& + \frac{426}{625} \tau_k^2 E^T K_1 B B^T K_1 E + \frac{9}{625} \tau_k^2 E^T K_2 B B^T K_2 E \\
& + \frac{72}{625} \tau_k^2 \left(E^T K_1 B B^T K_2 E + E^T K_2 B B^T K_1 E \right), \\
K_3 =& \tilde{K}_3 - \frac{17}{25} K_1, \\
\tilde{A}_k^T \tilde{K}_4 E + E^T \tilde{K}_4 \tilde{A}_k =& + \frac{981}{125} E^T K_1 E + \frac{177}{125} E^T K_2 E + \frac{1}{5} E^T K_3 E, \\
K_4 =& \tilde{K}_4 + K_3,
\end{aligned}
\tag{A.3}
$$

where the Lyapunov operator on the left hand side is given by
$\tilde{A}_k := \frac{1}{2}\left(\tau_k (A - B B^T X_k E) - E\right)$.

THESES

1. This thesis is concerned with the development of numerical methods for large-scale linear time-varying control systems. A finite-time optimal control problem with application to inverse problems and model order reduction for linear time-varying systems are considered. Throughout these methods, the differential Riccati and Lyapunov equations are essential ingredients and thus their efficient solution is of particular interest.
2. The tracking optimal control problem is applied to an inverse heat conduction problem in order to reconstruct the thermal source responsible for the outcome of a certain set of measurements.
3. The reconstruction quality of the inverse problem depends on the number and placements of the taken measurements. Driven by the design of the performance index, the identified heat source is constructed to be zero or at least close to zero. Being a set of measure zero this does not influence the reconstruction quality in the $\mathcal{L}_2$ sense.
4. The balanced truncation MOR method and interpolatory MOR based on Krylov subspace methods for linear time-invariant systems are reviewed. Within the frameworks of switched linear systems and parametric MOR, respectively, these reduction techniques are applied to linear time-varying systems.
5. The performance of the switched linear systems procedure directly depends on the number of subsystems. The approximation quality of the resulting reduced-order model cannot overcome the accuracy of the full-order switched system.
6. Balanced truncation for LTV systems is investigated. The switched linear systems and PMOR approaches for LTV systems are compared to the BT ansatz for LTV systems. It is shown that the BT for LTV scheme based on time-varying truncation matrices outperforms the other methods with respect to the accuracy but is not compatible regarding the offline and online timings. Still, a modi-

fication using global truncation matrices, similar to the PMOR approach, can compete in terms of the online simulation timings.

7. A number of implicit ODE integration methods for the application to differential matrix equations are investigated. The rather new classes of implicit peer and Rosenbrock-type peer methods is adapted to the solution of matrix-valued ODEs. For the latter a reformulation avoiding a number of Jacobian applications is proposed. This modification is most powerful for autonomous ODEs.

8. The solution of differential Riccati and Lyapunov equations by the presented ODE solvers can be reduced to the solution of a number of algebraic Lyapunov equations. Thus, the ADI and Krylov subspace based Lyapunov solvers are reviewed. Two different low-rank versions of these iterative solvers are investigated.

9. All classical implicit ODE solvers of integration order $s \geq 2$ result in indefinite right hand sides for the Lyapunov equations to be solved in the innermost iteration.

10. The superposition of the right hand side, as well as the solution of the Lyapunov equations, arising inside the LRCF based time integration methods, by a sum of positive and negative definite summands is corrupted by cancellation effects. Moreover, the LRCF ODE solvers of integration order $s \geq 2$ dealing with complex factorizations of the symmetric indefinite right hand side and complex shifts at the same time cannot be applied straightforwardly.

11. The use of the classical low-rank factorization results in complex right hand side decompositions and therefore requires complex arithmetic, storage consumption and thus requires a rank-revealing column compression technique capable of handling complex data.

12. The symmetric indefinite low-rank decomposition of the indefinite right hand sides avoids complex data and arithmetic. Further, the factorization reveals redundant information and allows an a-priori analytic column compression. The theoretically predicted superiority of the symmetric indefinite low-rank decomposition is verified by different numerical experiments.

13. Given definite right hand sides for the Lyapunov equations, the computational cost of the column compression technique for the LRSIF based methods exceeds their LRCF counterpart. Thus, using first-order integration schemes applied to SISO systems, the LRCF based schemes are recommended to use.

14. The number of ADI steps required for the solution of the Lyapunov equations in the innermost iteration at a certain time step decreases for decreasing time step sizes, whereas the number of Newton steps is step size invariant.

BIBLIOGRAPHY

[1] J. Abels and P. Benner, *CAREX – a collection of benchmark examples for continuous-time algebraic Riccati equations (version 2.0)*, Working Note 1999-14, SLICOT, Nov. 1999. Available from www.slicot.org. 160

[2] H. Abou-Kandil, G. Freiling, V. Ionescu, and G. Jank, *Matrix Riccati Equations in Control and Systems Theory*, Birkhäuser, Basel, Switzerland, 2003. 12, 13, 14

[3] H. Amann and J. Escher, *Analysis. III*, Birkhäuser Verlag, Basel, 2009, https://doi.org/10.1007/978-3-7643-7480-8. Translated from the 2001 German original by Silvio Levy and Matthew Cargo. 8

[4] D. Amsallem and C. Farhat, *Interpolation method for the adaptation of reduced-order models to parameter changes and its application to aeroelasticity*, AIAA J., 46 (2008), pp. 1803–1813. 54, 61

[5] E. Anderson, Z. Bai, C. Bischof, J. Demmel, J. Dongarra, J. Du Croz, A. Greenbaum, S. Hammarling, A. McKenney, and D. Sorensen, *LAPACK Users' Guide*, SIAM, Philadelphia, PA, third ed., 1999. 150

[6] A. C. Antoulas, *Approximation of Large-Scale Dynamical Systems*, SIAM Publications, Philadelphia, PA, 2005. 2, 3, 11, 13, 27, 29, 30, 31

[7] A. C. Antoulas, *A new result on passivity preserving model reduction*, Systems Control Lett., 54 (2005), pp. 361–374. 31

[8] A. C. Antoulas, D. C. Sorensen, and S. Gugercin, *A survey of model reduction methods for large-scale systems*, Contemp. Math., 280 (2001), pp. 193–219. 2, 27

[9] A. C. Antoulas, D. C. Sorensen, and Y. Zhou, *On the decay rate of Hankel singular values and related issues*, Systems Control Lett., 46 (2002), pp. 323–342. 16, 175

[10] U. M. Ascher and L. R. Petzold, *Computer Methods for Ordinary Differential Equations and Differential-Algebraic Equations*, SIAM, Philadelphia, PA, 1998. 93

[11] M. Athans and P. L. Falb, *Optimal Control*, McGraw-Hill, New York, 1966. 11, 38

[12] M. Barrault, Y. Maday, N. C. Nguyen, and A. T. Patera, *An 'empirical interpolation' method: application to efficient reduced-basis discretization of partial differential equations*, C. R. Math. Acad. Sci. Paris, 339 (2004), pp. 667–672. 61

[13] R. H. Bartels and G. W. Stewart, *Solution of the matrix equation $AX + XB = C$: Algorithm 432*, Comm. ACM, 15 (1972), pp. 820–826. 16, 22

[14] U. Baur, C. A. Beattie, P. Benner, and S. Gugercin, *Interpolatory projection methods for parameterized model reduction*, SIAM J. Sci. Comput., 33 (2011), pp. 2489–2518. 31, 54, 61, 62, 176

[15] U. Baur and P. Benner, *Modellreduktion für parametrisierte Systeme durch balanciertes Abschneiden und Interpolation (Model Reduction for Parametric Systems Using Balanced Truncation and Interpolation)*, at-Automatisierungstechnik, 57 (2009), pp. 411–420. 61

[16] U. Baur, P. Benner, and L. Feng, *Model order reduction for linear and nonlinear systems: A system-theoretic perspective*, Arch. Comput. Methods Eng., 21 (2014), pp. 331–358, https://doi.org/10.1007/s11831-014-9111-2. 27

[17] B. Beckermann and A. Gryson, *Extremal rational functions on symmetric discrete sets and superlinear convergence of the ADI method*, Constr. Approx., 32 (2010), pp. 393–428, https://doi.org/10.1007/s00365-010-9087-6. 17

[18] P. Benner, *Solving large-scale control problems*, IEEE Control Syst. Mag., 14 (2004), pp. 44–59. 16

[19] P. Benner and T. Breiten, *Low rank methods for a class of generalized Lyapunov equations and related issues*, Numerische Mathematik, 124 (2013), pp. 441–470, https://doi.org/10.1007/s00211-013-0521-0. 175

[20] P. Benner and Z. Bujanović, *On the solution of large-scale algebraic Riccati equations by using low-dimensional invariant subspaces*, Linear Algebra Appl., 488 (2016), pp. 430–459, https://doi.org/10.1016/j.laa.2015.09.027. 175

[21] P. Benner, S. Gugercin, and K. Willcox, *A survey of model reduction methods for parametric systems*, SIAM Review, 57 (2015), pp. 483–531, https://doi.org/10.1137/130932715. 3, 61, 175

[22] P. Benner, M. Heinkenschloss, J. Saak, and H. K. Weichelt, *An inexact low-rank Newton-ADI method for large-scale algebraic Riccati equations*, Appl. Numer. Math., 108 (2016), pp. 125–142, https://doi.org/10.1016/j.apnum.2016.05.006. 179

[23] P. Benner, P. Kürschner, and J. Saak, *Efficient handling of complex shift parameters in the low-rank Cholesky factor ADI method*, Numer. Algorithms, 62 (2013), pp. 225–251, https://doi.org/10.1007/s11075-012-9569-7. 3, 16, 18, 19, 175

[24] P. Benner, P. Kürschner, and J. Saak, *An improved numerical method for balanced truncation for symmetric second order systems*, Math. Comput. Model. Dyn. Sys., 19 (2013), pp. 593–615, https://doi.org/10.1080/13873954.2013.794363. 3, 16,

19, 175

[25] P. Benner, P. Kürschner, and J. Saak, *Self-generating and efficient shift parameters in ADI methods for large Lyapunov and Sylvester equations*, Electron. Trans. Numer. Anal., 43 (2014), pp. 142–162. 3, 16, 20, 175

[26] P. Benner, J.-R. Li, and T. Penzl, *Numerical solution of large Lyapunov equations, Riccati equations, and linear-quadratic control problems*, Numer. Lin. Alg. Appl., 15 (2008), pp. 755–777. 16, 19

[27] P. Benner, R.-C. Li, and N. Truhar, *On the ADI method for Sylvester equations*, J. Comput. Appl. Math., 233 (2009), pp. 1035–1045. 16, 22

[28] P. Benner, V. Mehrmann, and D. C. Sorensen, *Dimension Reduction of Large-Scale Systems*, vol. 45 of Lect. Notes Comput. Sci. Eng., Springer-Verlag, 2005. 2, 3, 13, 27, 189, 190, 193

[29] P. Benner and H. Mena, *BDF methods for large-scale differential Riccati equations*, in Proc. 16th Intl. Symp. Mathematical Theory of Network and Systems, MTNS 2004, B. De Moor, B. Motmans, J. Willems, P. Van Dooren, and V. Blondel, eds., 2004. 12 p., available at http://www.mtns2004.be. 92

[30] P. Benner and H. Mena, *Rosenbrock methods for solving Riccati differential equations*, IEEE Trans. Autom. Control, 58 (2013), pp. 2950–2957. 3, 92, 94, 96, 108

[31] P. Benner and H. Mena, *Numerical solution of the infinite-dimensional LQR-problem and the associated differential Riccati equations*, J. Numer. Math., (2016). 3, 92, 108

[32] P. Benner and E. S. Quintana-Ortí, *Solving stable generalized Lyapunov equations with the matrix sign function*, Numer. Algorithms, 20 (1999), pp. 75–100. 16, 22

[33] P. Benner and E. S. Quintana-Ortí, *Model reduction based on spectral projection methods*, in Benner et al. [28], pp. 5–45. 16, 22

[34] P. Benner and J. Saak, *A semi-discretized heat transfer model for optimal cooling of steel profiles*, in Benner et al. [28], pp. 353–356. 84, 153

[35] P. Benner and J. Saak, *Numerical solution of large and sparse continuous time algebraic matrix Riccati and Lyapunov equations: a state of the art survey*, GAMM Mitteilungen, 36 (2013), pp. 32–52, https://doi.org/10.1002/gamm.201310003. 15, 93

[36] J. G. Blom, W. Hundsdorfer, E. J. Spee, and J. G. Verwer, *A second order Rosenbrock method applied to photochemical dispersion problems*, SIAM J. Sci. Comput., 20(4) (1999), pp. 1456–1480. 97

[37] B. A. Boley and J. H. Weiner, *Theory of thermal stresses*, John Wiley & Sons Inc., New York - London, 1960. 68

[38] M. Bollhöfer and A. Eppler, *Low-rank Cholesky factor Krylov subspace methods for generalized projected Lyapunov equations*, in System Reduction for Nanoscale

IC Design, P. Benner, ed., vol. 20 of Mathematics in Industry, Springer-Verlag, Berlin/Heidelberg, Germany, 2017, pp. 157–193, https://doi.org/10.1007/978-3-319-07236-4_5. 151

[39] D. Bonesky, S. Dahlke, P. Maass, and T. Raasch, *Adaptive wavelets methods and sparsity reconstruction for inverse heat conduction problems*, Adv. Comput. Math., 33 (2010), pp. 385–411. 34

[40] S. Boyd, L. El Ghaoui, E. Feron, and V. Balakrishnan, *Linear matrix inequalities in system and control theory*, vol. 15 of SIAM Studies in Applied Mathematics, Society for Industrial and Applied Mathematics (SIAM), Philadelphia, PA, 1994, https://doi.org/10.1137/1.9781611970777. 58

[41] T. Breiten, *Interpolatory Methods for Model Reduction of Large-Scale Dynamical Systems*, Dissertation, Otto-von-Guericke-Universität, Magdeburg, Germany, 2013. 30

[42] A. Bunse-Gerstner, D. Kubalinska, G. Vossen, and D. Wilczek, *h_2-norm optimal model reduction for large scale discrete dynamical MIMO systems*, J. Comput. Appl. Math., 233 (2010), pp. 1202–1216, https://doi.org/10.1016/j.cam.2008.12.029. 31

[43] J. M. Burgers, *Application of a model system to illustrate some points of the statistical theory of free turbulence*, Proc. Roy. Netherl. Academy of Sciences (Am- sterdam), 43 (1940), pp. 2–12. 80

[44] J. M. Burgers, *A mathematical model illustrating the theory of turbulence*, Adv. Appl. Mech., 1 (1948), pp. 171–199, https://doi.org/10.1016/S0065-2156(08)70100-5. 80

[45] J. C. Butcher, *On the convergence of numerical solutions to ordinary differential equations*, Math. Comp., 20 (1966), pp. 1–10. 99

[46] J. C. Butcher, *General linear methods: a survey*, Appl. Numer. Math., 1 (1985), pp. 273–284, https://doi.org/10.1016/0168-9274(85)90007-8. 99

[47] J. C. Butcher, *General linear methods*, Comput. Math. Appl., 31 (1996), pp. 105–112, https://doi.org/10.1016/0898-1221(95)00222-7. 99

[48] J. L. Casti, *Dynamical systems and their applications: linear theory.*, Mathematics in Science and Engineering, Academic Press, New York - London, first ed., 1977. 1

[49] J. L. Casti, *Linear Dynamical Systems*, Mathematics in Science and Engineering, Academic Press, New York, 1987. 1, 11

[50] Y. Chahlaoui and P. Van Dooren, *Estimating Gramians of large-scale time-varying systems*, in Proceedings of the 15th IFAC World Congress, 2002. 3, 54

[51] Y. Chahlaoui and P. Van Dooren, *Model reduction of time-varying systems*, in Benner et al. [28], pp. 131–148, https://doi.org/10.1007/3-540-27909-1_5. 3, 54

[52] C. Choi and A. J. Laub, *Efficient matrix-valued algorithms for solving stiff Riccati differential equations*, IEEE Trans. Autom. Control, 35 (1990), pp. 770–776. 3, 13

[53] M. Cruz Varona, M. Geuss, and B. Lohmann, *Zeitvariante parametrische Modellordnungsreduktion am Beispiel von Systemen mit wandernder Last*, in Methoden und Anwendungen der Regelungstechnik, B. Lohmann and G. Roppenecker, eds., Shaker-Verlag, 2015. 54

[54] C. F. Curtiss and J. O. Hirschfelder, *Integration of stiff equations*, Proc. Nat. Acad. Sci. U. S. A., 38 (1952), pp. 235–243. 92

[55] E. J. Davison and M. C. Maki, *The numerical solution of the matrix Riccati differential equation*, IEEE Trans. Autom. Control, 18 (1973), pp. 71–73. 3, 91

[56] C. De Villemagne and R. E. Skelton, *Model reduction using a projection formulation*, Internat. J. Control, 46 (1987), pp. 2141–2169. 30, 31

[57] K. Dekker and J. G. Verwer, *Stability of Runge-Kutta methods for stiff nonlinear differential equations*, Elsevier, Amsterdam, North-Holland, 1984. 94, 95, 97

[58] P. Deuflhard and F. Bornemann, *Scientific computing with ordinary differential equations*, vol. 42 of Texts in Applied Mathematics, Springer-Verlag, 2002, https://doi.org/10.1007/978-0-387-21582-2. Translated from the 1994 German original by Werner C. Rheinboldt. 93

[59] L. Dieci, *Numerical integration of the differential Riccati equation and some related issues*, SIAM J. Numer. Anal., 29 (1992), pp. 781–815. 3, 13, 92, 97

[60] V. Druskin, L. Knizhnerman, and V. Simoncini, *Analysis of the rational Krylov subspace and ADI methods for solving the Lyapunov equation*, SIAM J. Numer. Anal., 49 (2011), pp. 1875–1898. 3, 16

[61] V. Druskin and V. Simoncini, *Adaptive rational Krylov subspaces for large-scale dynamical systems*, Systems and Control Letters, 60 (2011), pp. 546–560. 3, 16, 23, 31, 175

[62] V. Druskin, V. Simoncini, and M. Zaslavsky, *Adaptive tangential interpolation in rational Krylov subspaces for MIMO dynamical systems*, SIAM J. Matrix Anal. Appl., 35 (2014), pp. 476–498, https://doi.org/10.1137/120898784. 16

[63] R. Eid, B. Salimbahrami, B. Lohmann, E. B. Rudnyi, and J. G. Korvink, *Parametric order reduction of proportionally damped second-order systems*, Sensors and Materials, Tokyo, 19 (2007), pp. 149–164. 60

[64] D. Elliott, *Bilinear Control Systems*, vol. 169 of Applied Mathematical Sciences, Springer-Verlag, 2009, https://doi.org/10.1023/b101451. 178

[65] H. W. Engl, M. Hanke, and A. Neubauer, *Regularization of Inverse Problems*, vol. 375 of Mathematics and its Applications, Kluwer Academic Publishers Group, Dordrecht, 1996, https://doi.org/10.1007/978-94-009-1740-8. 34, 35

[66] D. F. ENNS, *Model reduction with balanced realizations: An error bound and a frequency weighted generalization*, in Proc. 23rd IEEE Conf. Decision Contr., vol. 23, 1984, pp. 127–132. 29

[67] M. R. ESLAMI, R. B. HETNARSKI, J. IGNACZAK, N. NODA, N. SUMI, AND Y. TANIGAWA, *Theory of Elasticity and Thermal Stresses*, vol. 197 of Solid Mechanics and its Applications, Springer, Dordrecht, 2013, https://doi.org/10.1007/978-94-007-6356-2. 68

[68] M. FARHOOD, C. L. BECK, AND G. E. DULLERUD, *Model reduction of periodic systems: a lifting approach*, Automatica J. IFAC, 41 (2005), pp. 1085–1090, https://doi.org/10.1016/j.automatica.2005.01.008. 3, 54

[69] O. FARLE, V. HILL, P. INGELSTRÖM, AND R. DYCZIJ-EDLINGER, *Multi-parameter polynomial order reduction of linear finite element models*, Math. Comput. Model. Dyn. Syst., 14 (2008), pp. 421–434. 60

[70] F. FEITZINGER, T. HYLLA, AND E. W. SACHS, *Inexact Kleinman-Newton method for Riccati equations*, SIAM J. Matrix Anal. Appl., 31 (2009), pp. 272–288. 179

[71] L. FENG, *Parameter independent model order reduction*, Math. Comput. Simulation, 68 (2005), pp. 221–234. 60

[72] L. FENG AND P. BENNER, *A robust algorithm for parametric model order reduction*, Proc. Appl. Math. Mech., 7 (2008), pp. 1021501–1021502. 60

[73] L. FENG, E. B. RUDNYI, AND J. G. KORVINK, *Preserving the film coefficient as a parameter in the compact thermal model for fast electro-thermal simulation*, IEEE Trans. Comput.-Aided Design Integr. Circuits Syst., 24 (2005), pp. 1838–1847. 60

[74] M. FISCHER AND P. EBERHARD, *Simulation of moving loads in elastic multibody systems with parametric model reduction techniques*, Arch. Mech. Eng., 61 (2014), pp. 209–216. 3, 54

[75] M. FISCHER AND P. EBERHARD, *Application of parametric model reduction with matrix interpolation for simulation of moving loads in elastic multibody systems*, Adv. Comput. Math., 41 (2015), pp. 1049–1072. 3, 54

[76] B. A. FRANCIS, *A Course In $\mathcal{H}_\infty$ Control Theory*, vol. 88 of Lecture Notes in Control and Information Sciences, Springer-Verlag, Berlin, 1987, https://doi.org/10.1007/BFb0007371. 2

[77] F. FREITAS, J. ROMMES, AND N. MARTINS, *Gramian-based reduction method applied to large sparse power system descriptor models*, IEEE Trans. Power Syst., 23 (2008), pp. 1258–1270. 70

[78] R. W. FREUND, *Model reduction methods based on Krylov subspaces*, Acta Numer., 12 (2003), pp. 267–319. 2, 27

[79] K. GALLIVAN, A. VANDENDORPE, AND P. VAN DOOREN, *Model reduction of MIMO*

systems via tangential interpolation, SIAM J. Matrix Anal. Appl., 26 (2004), pp. 328–349. 30

[80] J. D. Gardiner, A. J. Laub, J. J. Amato, and C. B. Moler, *Solution of the Sylvester matrix equation $AXB + CXD = E$*, ACM Trans. Math. Software, 18 (1992), pp. 223–231. 16

[81] M. Geuss and K. J. Diepold, *An approach for stability-preserving model order reduction for switched linear systems based on individual subspaces*, in Methoden und Anwendungen der Regelungstechnik, G. Roppenecker and B. Lohmann, eds., Shaker-Verlag, Aachen, Sept. 2013. 57

[82] K. Glover, *All optimal Hankel-norm approximations of linear multivariable systems and their L^∞-error norms*, Internat. J. Control, 39 (1984), pp. 1115–1193. 28

[83] S. K. Godunov, *Ordinary Differential Equations with Constant Coefficient*, vol. 169 of Translations of Mathematical Monographs, AMS, Providence, RI, 1997. 80

[84] G. H. Golub and C. F. Van Loan, *Matrix Computations*, Johns Hopkins University Press, Baltimore, third ed., 1996. 23, 132

[85] L. Grasedyck, *Existence of a low rank or H-matrix approximant to the solution of a Sylvester equation*, Numer. Lin. Alg. Appl., 11 (2004), pp. 371–389. 16

[86] M. Green and D. J. N. Limebeer, *Linear Robust Control*, Prentice-Hall, Englewood Cliffs, NJ, 1995. 2

[87] M. A. Grepl, Y. Maday, N. C. Nguyen, and A. T. Patera, *Efficient reduced-basis treatment of nonaffine and nonlinear partial differential equations*, ESAIM: Math. Model. Numer. Anal., 41 (2007), pp. 575–605. 61

[88] E. J. Grimme, *Krylov projection methods for model reduction*, Ph.D. Thesis, Univ. of Illinois at Urbana-Champaign, USA, 1997. 30, 31

[89] K. Grossmann, C. Städel, A. Galant, and A. Mühl, *Berechnung von Temperaturfeldern an Werkzeugmaschinen*, Zeitschrift für Wirtschaftlichen Fabrikbetrieb, 107 (2012), pp. 452–456. 66, 71

[90] S. Gugercin, A. C. Antoulas, and C. A. Beattie, *$\mathcal{H}_2$ model reduction for large-scale dynamical systems*, SIAM J. Matrix Anal. Appl., 30 (2008), pp. 609–638. 31, 61

[91] S. Gugercin and J.-R. Li, *Smith-type methods for balanced truncation of large systems*, in Benner et al. [28], pp. 49–82. 30

[92] B. Haasdonk and M. Ohlberger, *Efficient reduced models for parametrized dynamical systems by offline/online decomposition*, in Proc. MATHMOD 2009, 6th Vienna International Conference on Mathematical Modelling, 2009. 62

[93] B. Haasdonk, M. Ohlberger, and G. Rozza, *A reduced basis method for evolution schemes with parameter-dependent explicit operators*, Electron. Trans. Numer. Anal., 32 (2008), pp. 145–168. 61

[94] J. Hadamard, *Sur les problèmes aux dérivés partielles et leur signification physique*, Princeton University Bulletin, 13 (1902), pp. 49–52. 34, 35

[95] E. Hairer, S. P. Nørsett, and G. Wanner, *Solving Ordinary Differential Equations I-Nonstiff Problems*, Springer Series in Computational Mathematics, Springer-Verlag, New York, 2000. 92, 93, 94

[96] E. Hairer and G. Wanner, *Solving Ordinary Differential Equations II - Stiff and Differential-Algebraic Problems*, Springer Series in Computational Mathematics, Springer-Verlag, second ed., 2002. 92, 93, 94, 96, 102

[97] S. J. Hammarling, *Numerical solution of the stable, non-negative definite Lyapunov equation*, IMA J. Numer. Anal., 2 (1982), pp. 303–323. 16, 22

[98] E. Hansen and T. Stillfjord, *Convergence analysis for splitting of the abstract differential riccati equation*, SIAM J. Numer. Anal., 52 (2014), pp. 3128–3139, https://doi.org/10.1137/130935501. 4, 91

[99] J. Harnard, P. Winternitz, and R. L. Anderson, *Superposition principles for matrix Riccati equations*, J. Math. Phys., 24 (1983), pp. 1062–1072. 3, 91

[100] S. Hein, *MPC-LQG-Based Optimal Control of Parabolic PDEs*, Dissertation, TU Chemnitz, February 2009. Available from http://archiv.tu-chemnitz.de/pub/2010/0013. 80

[101] D. Hinrichsen and A. J. Pritchard, *Mathematical Systems Theory I*, Springer-Verlag, Berlin, 2005. 11

[102] B. Hofmann, *Mathematik inverser Probleme*, Teubner, Stuttgart-Leipzig, Germany, 1999. In German. 35

[103] B. Hofmann and S. Kindermann, *On the degree of ill-posedness for linear problems with non-compact operators*, Methods of Applications and Analysis, (2010), pp. 445–462. 35

[104] R. A. Horn and C. R. Johnson, *Matrix Analysis*, Cambridge University Press, Cambridge, 1985. 150

[105] M.-S. Hossain and P. Benner, *Projection-based model reduction for LTV descriptor system using multipoint Krylov-subspace projectors*, Proc. Appl. Math. Mech., 8 (2008), pp. 10081–10084. 2, 54

[106] A. Ichikawa and H. Katayama, *Remarks on the time-varying H_∞ Riccati equations*, Systems Control Lett., 37 (1999), pp. 335–345. 12

[107] A. Isidori, *Nonlinear Control Systems*, Communications and Control Engineering, Springer-Verlag, 3 ed., 1995, https://doi.org/10.1007/978-1-84628-615-5. 178

[108] O. L. R. Jacobs, *Introduction to Control Theory*, Oxford Science Publications, Oxford, UK, 2nd ed., 1993. 12

[109] I. M. Jaimoukha and E. M. Kasenally, *Krylov subspace methods for solving large Lyapunov equations*, SIAM J. Numer. Anal., 31 (1994), pp. 227–251. 16

[110] B. Kågström and P. Poromaa, *LAPACK-style Algorithms and Software for Solving the Generalized Sylvester Equation and Estimating the Separation Between Regular Matrix Pairs*, ACM Trans. Math. Software, 22 (1996), pp. 78–103, https://doi.org/10.1145/225545.225552. 16

[111] B. Kågström and L. Westin, *Generalized Schur methods with condition estimators for solving the generalized Sylvester equation*, IEEE Trans. Automat. Control, 34 (1989), pp. 745–751, https://doi.org/10.1109/9.29404. 16

[112] T. Kailath, *Linear Systems*, Prentice-Hall, Englewood Cliffs, NJ, 1980. 1, 8, 58

[113] R. E. Kalman, *Lyapunov functions for the problem of Lur'e in automatic control*, Proc. Nat. Acad. Sci. U.S.A., 49 (1963), pp. 201–205. 58

[114] C. Kenney and R. B. Leipnik, *Numerical integration of the differential matrix Riccati equation*, IEEE Trans. Autom. Control, 30 (1985), pp. 962–970. 3, 91

[115] S. Kindermann and A. Leitão, *On regularization methods based on dynamic programming techniques*, Appl. Anal., 86 (2007), pp. 611–632, https://doi.org/10.1080/00036810701354953. 36

[116] S. Kindermann and C. Navasca, *Optimal control as a regularization method for ill-posed problems*, J. Inverse Ill-Posed Probl., 14 (2006), pp. 685–703. 36

[117] D. L. Kleinman, *On an iterative technique for Riccati equation computations*, IEEE Trans. Autom. Control, 13 (1968), pp. 114–115. 15

[118] H. W. Knobloch and H. Kwakernaak, *Lineare Kontrolltheorie*, Springer-Verlag, Berlin, 1985. In German. 13

[119] M. Köhler, N. Lang, and J. Saak, *Solving differential matrix equations using parareal*, Proc. Appl. Math. Mech., 16 (2016), pp. 847–848, https://doi.org/10.1002/pamm.201610412. 153, 178

[120] M. Köhler and J. Saak, *Efficiency Improving Implementation Techniques for Large Scale Matrix Equation Solvers*, Chemnitz Scientific Computing Prep. CSC 09-10, TU Chemnitz, 2009. 16, 22

[121] M. Köhler and J. Saak, *On BLAS level-3 implementations of common solvers for (quasi-) triangular generalized Lyapunov equations*, ACM Trans. Math. Software, 43 (2016), pp. 3:1–3:23, https://doi.org/10.1145/2850415. 16, 22

[122] M. Köhler and J. Saak, *On GPU acceleration of common solvers for (quasi-) triangular generalized Lyapunov equations*, Parallel Comput., (2016), https://doi.org/10.1016/j.parco.2016.05.010. 16, 22

[123] D. D. Kosambi, *Statistics in function space*, J. Ind. Math. Soc., 7 (1943), pp. 76–88. 3

[124] A. Koskela and H. Mena, *A structure preserving Krylov subspace method for large scale differential Riccati equations*, Tech. Report 1705.07507, Cornell University, may 2017, https://arxiv.org/abs/1705.07507. 91

[125] P. Kunkel and V. Mehrmann, *Differential-Algebraic Equations: Analysis and Numerical Solution*, Textbooks in Mathematics, EMS Publishing House, Zürich, Switzerland, 2006. 3, 91

[126] P. Kürschner, *Efficient Low-Rank Solution of Large-Scale Matrix Equations*, Dissertation, Otto-von-Guericke-Universität, Apr. 2016, http://hdl.handle.net/11858/00-001M-0000-0029-CE18-2. 3, 16, 17, 18, 19, 20, 130, 175

[127] D. G. Lainiotis, *Generalized Chandrasekhar algorithms: Time-varying models*, IEEE Trans. Automat. Control, 21 (1976), pp. 728–732. 91

[128] S. Lall and C. Beck, *Error-bounds for balanced model-reduction of linear time-varying systems*, IEEE Trans. Automat. Control, 48 (2003), pp. 946–956, https://doi.org/10.1109/TAC.2003.812779. 3, 54

[129] P. K. Lamm, *Future-sequential regularization methods for ill-posed Volterra equations*, J. of Math. Anal. Appl., (1996), pp. 469–494. 35

[130] P. Lancaster and L. Rodman, *Algebraic Riccati Equations*, Oxford University Press, Oxford, UK, 1995. 14, 15

[131] N. Lang, H. Mena, and J. Saak, *An LDL^T factorization based ADI algorithm for solving large scale differential matrix equations*, Proc. Appl. Math. Mech., 14 (2014), pp. 827–828, https://doi.org/10.1002/pamm.201410394. 16

[132] N. Lang, H. Mena, and J. Saak, *On the benefits of the LDL^T factorization for large-scale differential matrix equation solvers*, Linear Algebra Appl., 480 (2015), pp. 44–71, https://doi.org/10.1016/j.laa.2015.04.006. 5, 16

[133] N. Lang, J. Saak, and P. Benner, *Model order reduction for systems with moving loads*, at-Automatisierungstechnik, 62 (2014), pp. 512–522. 5

[134] N. Lang, J. Saak, P. Benner, S. Ihlenfeldt, S. Nestmann, and K. Schädlich, *Towards the identification of heat induction in chip removing processes via an optimal control approach*, Production Engineering, 9 (2015), pp. 343–349, https://doi.org/10.1007/s11740-015-0608-9. 5

[135] N. Lang, J. Saak, and T. Stykel, *Balanced truncation model reduction for linear time-varying systems*, Math. Comput. Model. Dyn. Syst., 22 (2016), pp. 267–281, https://doi.org/10.1080/13873954.2016.1198386. 5

[136] A. J. Laub, *Schur techniques for Riccati differential equations*, in Feedback Control of Linear and Nonlinear Systems, D. Hinrichsen and A. Isidori, eds., Springer-Verlag, New York, 1982, pp. 165–174. 3, 91

[137] A. J. Laub, M. T. Heath, C. C. Paige, and R. C. Ward, *Computation of system balanc-*

ing transformations and other applications of simultaneous diagonalization algorithms, IEEE Trans. Autom. Control, 32 (1987), pp. 115–122. 28

[138] A. T.-M. Leung and R. Khazaka, *Parametric model order reduction technique for design optimization*, Circuits and Systems, 2005. ISCAS 2005. IEEE International Symposium on, 2 (2005), pp. 1290–1293, https://doi.org/10.1109/ISCAS.2005.1464831. 60

[139] J.-R. Li, *Model Reduction of Large Linear Systems via Low Rank System Gramians*, Ph.D. Thesis, Massachusettes Institute of Technology, Sept. 2000. 16, 18

[140] J.-R. Li and J. White, *Low rank solution of Lyapunov equations*, SIAM J. Matrix Anal. Appl., 24 (2002), pp. 260–280. 16, 18, 19

[141] D. Liberzon, *Switching in Systems and Control*, Springer-Verlag, New York, 2003. 55, 56, 175

[142] J.-L. Lions, Y. Maday, and G. Turinici, *Résolution d'EDP par un schéma en temps "pararéel"*, Comptes Rendus de l'Académie des Sciences. Série I. Mathématique, 332 (2001), pp. 661–668, https://doi.org/10.1016/S0764-4442(00)01793-6. 178

[143] A. Locatelli, *Optimal Control: An Introduction*, Birkhäuser, Basel, Switzerland, 2001. 2, 13, 24, 25, 26, 38, 51, 175

[144] S. Longhi and G. Orlando, *Balanced reduction of linear periodic systems*, Kybernetika (Prague), 35 (1999), pp. 737–751. 3, 54

[145] A. K. Louis, *Inverse und schlecht gestellte Probleme*, Teubner, 1989. 34

[146] D. G. Luenberger, *Introduction to Dynamic Systems. Theory, Models, and Applications.*, John Wiley & Sons., New York etc., first ed., 1979. 1, 2, 11

[147] J. Macki and A. Strauss, *Introduction to Optimal Control Theory*, Springer-Verlag, 1982. 2

[148] V. Mehrmann, *The Autonomous Linear Quadratic Control Problem, Theory and Numerical Solution*, no. 163 in Lecture Notes in Control and Information Sciences, Springer-Verlag, Heidelberg, July 1991. 3, 91

[149] H. Mena, *Numerical Solution of Differential Riccati Equations Arising in Optimal Control Problems for Parabolic Partial Differential Equations*, Ph.D. Thesis, Escuela Politecnica Nacional, 2007. 3, 13, 91, 92, 94, 96, 108, 152

[150] R. R. Mohler, *Bilinear Control Processes: With Applications to Engineering, Ecology and Medicine*, Academic Press, Inc., Orlando, FL, USA, 1973. 178

[151] N. Monshizadeh, H. L. Trentelman, and M. K. Çamlibel, *Simultaneous balancing and model reduction of switched linear systems*, in 50th IEEE Conference on Decision and Control and European Control Conference (CDC-ECC), Orlando, Florida, 2011, pp. 6552–6557, https://doi.org/10.1109/CDC.2011.6160263. 56

[152] B. C. Moore, *Principal component analysis in linear systems: controllability, observability, and model reduction*, IEEE Trans. Autom. Control, AC-26 (1981), pp. 17–32. 27, 29

[153] C. Mullis and R. A. Roberts, *Synthesis of minimum roundoff noise fixed point digital filters*, IEEE Trans. Circuits and Systems, CAS-23 (1976), pp. 551–562. 27

[154] R. Neugebauer, W. G. Drossel, S. Ihlenfeldt, and C. Richter, *Thermal interactions between the process and workpiece.*, Procedia CIRP, (2012), pp. 63–66, https://doi.org/10.1016/j.procir.2012.10.012. 33

[155] H. Panzer, J. Mohring, R. Eid, and B. Lohmann, *Parametric model order reduction by matrix interpolation*, at-Automatisierungstechnik, 58 (2010), pp. 475–484. 54, 61

[156] D. Peaceman and H. Rachford, *The numerical solution of elliptic and parabolic differential equations*, J. Soc. Indust. Appl. Math., 3 (1955), pp. 28–41. 17

[157] T. Penzl, *Numerical solution of generalized Lyapunov equations*, Adv. Comp. Math., 8 (1997), pp. 33–48. 16

[158] T. Penzl, *A cyclic low rank Smith method for large sparse Lyapunov equations*, SIAM J. Sci. Comput., 21 (2000), pp. 1401–1418. 16, 19, 20

[159] T. Penzl, *Eigenvalue decay bounds for solutions of Lyapunov equations: the symmetric case*, Systems Control Lett., 40 (2000), pp. 139–144. 16, 19, 175

[160] I. R. Petersen, V. A. Ugrinovskii, and A. V. Savkin, *Robust Control Design Using H^∞ Methods*, Springer-Verlag, London, UK, 2000. 12

[161] J. R. Phillips, *Projection-based approaches for model reduction of weakly nonlinear, time-varying systems*, IEEE Trans. Comput.-Aided Design Integr. Circuits Syst., 22 (2003), pp. 171–187. 2, 54

[162] H. Podhaisky, R. Weiner, and B. A. Schmitt, *Rosenbrock-type 'peer' two-step methods*, Appl. Numer. Math., 53 (2005), pp. 409–420, https://doi.org/10.1016/j.apnum.2004.08.021. 99, 101

[163] H. Podhaisky, R. Weiner, and B. A. Schmitt, *Linearly-implicit two-step methods and their implementation in Nordsieck form*, Appl. Numer. Math., 56 (2006), pp. 374–387, https://doi.org/10.1016/j.apnum.2005.04.024. 99

[164] W. T. Reid, *Riccati Differential Equations*, Academic Press, New York, 1972. 13

[165] H. H. Rosenbrock, *Some general implicit processes for the numerical solution of differential equations*, Comput. J., 5 (1962/1963), pp. 329–330. 94

[166] W. J. Rugh, *Nonlinear System Theory: the Volterra/Wiener approach*, Johns Hopkins series in information sciences and systems, Johns Hopkins University Press, 1981, https://books.google.de/books?id=XvRQAAAAMAAJ. 178

[167] Y. Saad, *Numerical solution of large Lyapunov equation*, in Signal Processing, Scattering, Operator Theory and Numerical Methods, M. A. Kaashoek, J. H. van Schuppen, and A. C. M. Ran, eds., Birkhäuser, 1990, pp. 503–511. 16, 22

[168] J. Saak, *Efficient Numerical Solution of Large Scale Algebraic Matrix Equations in PDE Control and Model Order Reduction*, Dissertation, TU Chemnitz, July 2009. available from http://nbn-resolving.de/urn:nbn:de:bsz:ch1-200901642. 16, 84, 153

[169] A. Saberi, P. Sannuti, and B. M. Chen, *H_2 Optimal Control*, Prentice-Hall International Series in Systems and Control Engineering, Prentice-Hall, Hertfordshire, UK, 1995. 2

[170] J. Sabino, *Solution of Large-Scale Lyapunov Equations via the Block Modified Smith Method*, Ph.D. Thesis, Rice University, Houston, Texas, June 2007. available from: http://www.caam.rice.edu/tech_reports/2006/TR06-08.pdf. 16, 20

[171] H. Sandberg and A. Rantzer, *Balanced truncation of linear time-varying systems*, IEEE Trans. Automat. Control, 49 (2004), pp. 217–229. 3, 54, 63, 175, 178

[172] R. L. Schilling, *Measures, Integrals and Martingales*, Cambridge University Press, New York, 2005, https://doi.org/10.1017/CBO9780511810886. 8

[173] S. Schindler, M. Zimmermann, J. Aurich, and P. Steinmann, *Finite element model to calculate the thermal expansions of the tool and the workpiece in dry turning.*, Procedia CIRP, (2014), pp. 535–540. 33

[174] S. Schindler, M. Zimmermann, J. Aurich, and P. Steinmann, *Thermo-elastic deformations of the workpiece when dry turning aluminum alloys - a finite element model to predict thermal effects in the workpiece.*, CIRP Journal of Manufacturing Science and Technology, (2014), pp. 233–245, https://doi.org/10.1016/j.cirpj.2014.04.006. 33

[175] B. A. Schmitt and R. Weiner, *Parallel two-step W-methods with peer variables*, SIAM J. Numer. Anal., 42 (2004), pp. 265–282 (electronic), https://doi.org/10.1137/S0036142902411057. 92, 99

[176] B. A. Schmitt, R. Weiner, and K. Erdmann, *Implicit parallel peer methods for stiff initial value problems*, Appl. Numer. Math., 53 (2005), pp. 457–470, https://doi.org/10.1016/j.apnum.2004.08.019. 99

[177] B. A. Schmitt, R. Weiner, and H. Podhaisky, *Multi-implicit peer two-step W-methods for parallel time integration*, BIT, 45 (2005), pp. 197–217, https://doi.org/10.1007/s10543-005-2635-y. 99

[178] M. Schweinoch, R. Joliet, and P. Kersting, *Predicting thermal loading in NC milling processes.*, Production Engineering, 9 (2014), pp. 179–186, https://doi.org/10.1007/s11740-014-0598-z. 33

[179] H. R. Shaker, *Model Reduction of Hybrid Systems*, Ph.D. Thesis, Aalborg Univer-

sity, Aalborg, Denmark, 2010. 56

[180] H. R. Shaker and R. Wisniewski, *Generalised Gramian framework for model/controller order reduction of switched systems*, Int. J. Syst. Sci., 42 (2011), pp. 1277–1291. 56

[181] H. R. Shaker and R. Wisniewski, *Model reduction of switched systems based on switching generalized Gramians*, Int. J. Innov. Comput. I., 8 (2012), pp. 5025–5044. 56

[182] L. F. Shampine, *Implementation of Rosenbrock methods*, ACM Transactions on Mathematical Software, 8 (1982), pp. 93–103. 153, 183

[183] S. Shokoohi, L. Silverman, and P. Van Dooren, *Linear time-variable systems: balancing and model reduction*, IEEE Trans. Automat. Control, 28 (1983), pp. 810–822. 3, 54, 63, 64, 175

[184] V. Sima, *Algorithms for Linear-Quadratic Optimization*, vol. 200 of Pure and Applied Mathematics, Marcel Dekker, Inc., New York, NY, 1996. 2

[185] V. Simoncini, *A new iterative method for solving large-scale Lyapunov matrix equations*, SIAM J. Sci. Comput., 29 (2007), pp. 1268–1288. 3, 16, 23, 175

[186] V. Simoncini, *Computational methods for linear matrix equations*, SIAM Review, 58 (2016), pp. 377–441, https://doi.org/10.1137/130912839. 15, 93

[187] B. Soleimani and R. Weiner, *A class of implicit peer methods for stiff systems*, Journal of Computational and Applied Mathematics, 316 (2017), pp. 358 – 368, https://doi.org/https://doi.org/10.1016/j.cam.2016.06.014. Selected Papers from NUMDIFF-14. 99, 100

[188] E. D. Sontag, *Mathematical Control Theory*, Springer-Verlag, New York, NY, 2nd ed., 1998. 11

[189] G. Starke, *Optimal alternating directions implicit parameters for nonsymmetric systems of linear equations*, SIAM J. Numer. Anal., 28 (1991), pp. 1431–1445. 20

[190] T. Stillfjord, *Low-rank second-order splitting of large-scale differential Riccati equations*, IEEE Trans. Autom. Control, 61 (2015), pp. 2791–2796, https://doi.org/10.1109/TAC.2015.2398889. 4, 91, 179

[191] T. Stillfjord, *Adaptive high-order splitting schemes for large-scale differential Riccati equations*, arxiv e-prints 1612.00677, Cornell University, Dec. 2016, https://arxiv.org/abs/1612.00677. 179

[192] T. Stillfjord, *Adaptive high-order splitting schemes for large-scale differential riccati equations*, Numerical Algorithms, (2017), https://doi.org/10.1007/s11075-017-0416-8. 4, 91

[193] K. Strehmel, R. Weiner, and H. Podhaisky, *Numerik gewöhnlicher Differentialgleichungen*, Vieweg+Teubner-Verlag, 2nd ed., 2012, https://doi.org/10.1007/

978-3-8348-2263-5. 99

[194] T. Stykel and V. Simoncini, *Krylov subspace methods for projected Lyapunov equations*, Appl. Numer. Math., 62 (2012), pp. 35–50, https://doi.org/10.1016/j.apnum.2011.09.007. 3, 16

[195] T. Stykel and A. Vasilyev, *A two-step model reduction approach for mechanical systems with moving loads*, J. Comput. Appl. Math., 297 (2016), pp. 85–97, https://doi.org/10.1016/j.cam.2015.11.014. 2, 54

[196] A. N. Tikhonov, *Solution of incorrectly formulated problems and the regularization method*, in Soviet Math. Dokl, vol. 4, 1963, pp. 1035–1038. 35

[197] M. S. Tombs and I. Postlethwaite, *Truncated balanced realization of a stable non-minimal state-space system*, Internat. J. Control, 46 (1987), pp. 1319–1330. 28

[198] N. Truhar and K. Veselić, *Bounds on the trace of a solution to the Lyapunov equation with a general stable matrix*, Systems Control Lett., 56 (2007), pp. 493–503, https://doi.org/10.1016/j.sysconle.2007.02.003. 16, 175

[199] A. van der Schaft and H. Schumacher, *An Introduction to Hybrid Dynamical Systems*, vol. 251 of Lecture Notes in Control and Information Sciences, Springer-Verlag London, Ltd., London, 2000, https://doi.org/10.1007/BFb0109998. 55

[200] P. Van Dooren, K. Gallivan, and P.-A. Absil, *$\mathcal{H}_2$-optimal model reduction of MIMO systems*, Appl. Math. Lett., 21 (2008), pp. 1267–1273, https://doi.org/10.1016/j.aml.2007.09.015. 31

[201] A. Varga, *Balanced truncation model reduction of periodic systems*, in Proc. 39th IEEE Conference on Decision and Control, Sydney, Australia, vol. 3, 2000, pp. 2379–2384. 3, 54

[202] E. I. Verriest and T. Kailath, *On generalized balanced realizations*, IEEE Trans. Automat. Control, 28 (1983), pp. 833–844. 3, 54, 63, 175

[203] E. L. Wachspress, *The ADI Model Problem*, Springer-Verlag, 2013, https://doi.org/10.1007/978-1-4614-5122-8. 17, 20

[204] H. K. Weichelt, *Numerical Aspects of Flow Stabilization by Riccati Feedback*, Dissertation, Otto-von-Guericke-Universität, 2016, http://edoc2.bibliothek.uni-halle.de/hs/content/titleinfo/62076. 179

[205] D. S. Weile, E. Michielssen, E. Grimme, and K. Gallivan, *A method for generating rational interpolant reduced order models of two-parameter linear systems*, Appl. Math. Lett., 12 (1999), pp. 93–102. 60

[206] K. Wulff, *Quadratic and Non-Quadratic Stability Criteria for Switched Linear Systems*, Ph.D. Thesis, National University of Ireland, Maynooth, Ireland, 2004. 55

[207] A. Yousuff and R. E. Skelton, *Covariance equivalent realizations with application*

to model reduction of large-scale systems, in Control and Dynamic Systems, Vol. 22, Academic Press, Orlando, FL, 1985, pp. 273–348. 30

[208] A. Yousuff, D. A. Wagie, and R. E. Skelton, *Linear system approximation via covariance equivalent realizations*, J. Math. Anal. Appl., 106 (1985), pp. 91–115, https://doi.org/10.1016/0022-247X(85)90133-7. 30